Conservation in the built environment raises fundar
have been debated for centuries – what is worth preserving, how is it possible and why is it important?

This book takes a modern approach to the meaning of a heritage structure and its conservation. The historical evolution of conservation is briefly addressed, considering prominent individuals and cases; along with the history of construction, focusing on materials and related structural elements, with insight on the sizing rules adopted by masons. This explains structural decisions made during the construction process and allows comparison of scientific theories from the 18th century to modern understanding of limit analysis. Damage and collapse mechanisms for masonry construction, as the most widespread structural form for historical buildings, are described. Excess permanent loading and settlement are differentiated from environmental and anthropogenic actions such as earthquake or incorrect intervention.

The team of authors brings together unique expertise, with high-level research and leading practice with archetypical cases from around the world. The book addresses the history of conservation by exploring materials and structures and the history of construction and damage, so it is of value to students and professionals in civil engineering and architecture, as well as archaeologists and art historians.

Dr. Pere Roca is a professor at the School of Civil Engineering of Barcelona, Technical University of Catalonia (UPC). He has participated in a large number of national and international research projects and has been a consultant in over a hundred case studies on existing and historical structures. He is a co-editor of the *International Journal of Architectural Heritage*.

Dr. Paulo B. Lourenço is a professor at the Department of Civil Engineering, University of Minho, Guimarães, Portugal. He has been the co-head of the Institute in Sustainability and Innovation in Structural Engineering since 2007 and the co-head of the Institute for Bio-Sustainability since 2013. He is also a co-editor of the *International Journal of Architectural Heritage*.

Dr. Angelo Gaetani is a postdoctoral research fellow at the Department of Civil Engineering, University of Minho in Guimarães (Portugal). He worked as a professional structural engineer and consultant in Rome (Italy) until 2012.

Historic Construction and Conservation

Historic Construction and Conservation
Materials, Systems and Damage

Pere Roca, Paulo B. Lourenço and
Angelo Gaetani

Routledge
Taylor & Francis Group
New York London

Cover photo: © Nuno Mendes, Carmo Convent, Lisbon, Portugal

First published 2019
by Routledge
605 Third Avenue, New York, NY 10017

and by Routledge
2 Park Square, Milton Park, Abingdon, Oxon, OX14 4RN

First issued in paperback 2021

Routledge is an imprint of the Taylor & Francis Group, an informa business

© 2020 Taylor & Francis

The right of Pere Roca, Paulo B. Lourenço and Angelo Gaetani to be identified as authors of this work has been asserted by them in accordance with sections 77 and 78 of the Copyright, Designs and Patents Act 1988.

Publisher's Note
The publisher has gone to great lengths to ensure the quality of this reprint but points out that some imperfections in the original copies may be apparent.

ISBN 13: 978-1-03-209023-8 (pbk)
ISBN 13: 978-0-367-14574-3 (hbk)

Contents

List of Figures

List of Tables

Preface

What does *conservation* mean? Conservation may be defined as the '*careful preservation and protection of something... to prevent exploitation, destruction, or neglect*' (from Merriam-Webster dictionary). While the definition is obviously correct, as far as built cultural heritage is concerned, a few questions remain unanswered: what is worth to be preserved, how to make this possible and why it is important? To find answers, many scholars, experts and artists have debated the subject for centuries, and the discussion remains open.

But why such a complexity? The main source of complication is given by the history of human activities. Each monument, built space or infrastructure represents a place where populations of the past have lived, traded, prayed or fought. All these activities are engraved, sometimes hidden, upon a heritage building, and high-specialized expertise is requested to interpret this information and to reveal it to the community. In this regard, Andrea Carandini, President of FAI (*Fondo Ambiente Italiano*, or the National Trust of Italy) for the recent re-opening of the 1,000-year-old Church within the Abbey of Saint Mary in Cerrate (Lecce, Italy) stated:

> A monument is not a static unit but a stream of human and natural actions that never stops flowing, sometimes impetuously, in various branches. Giving value to a monument means knowing the meanders of this human flowing, discovering them, understanding their beauty and history, and recount them – perhaps after decades of decadence and wrong actions – to find lost origins and to revive a local community today.

From the cited words, it is possible to individuate one of the most important aspects of conservation: *authenticity*. This was the subject of the 'Conference on authenticity in relation to the world heritage convention' held in Nara, Japan (ICOMOS, 1994). A portion of the final document reads:

> In a world that is increasingly subject to the forces of globalization and homogenization, and in a world in which the search for cultural identity is sometimes pursued through aggressive

nationalism and the suppression of the cultures of minorities, the essential contribution made by the consideration of authenticity in conservation practice is to clarify and illuminate the collective memory of humanity.

Conservation is thus also a synonym of efforts, responsibility and respect, often linked to high-technical expertise. A single book cannot contain all the information requested to carry on a proper conservation project. The aspects involved are extensive and subjective, meaning that they are, to some extent, left to the sensitivity and experience of the persons involved, on a case-by-case basis. The present book aims to address the values of conservation, to highlight historical examples of misinterpretations and successes, to provide fundamental details on historical construction materials and techniques, as well as to describe the main damages for historical buildings. Without claiming to fully treat these topics, for which an extensive literature may be available in some cases, the present work provides a stimulus for the reader on the importance of investigation at any level and in any field. Each of the stakeholders involved in the project (engineer, architect, archaeologist, historian, artist, workman or other) needs to understand that a profound knowledge of the cultural heritage building is at the base of respectful conservation activities. According to ICOMOS (1994),

> Knowledge and understanding of these various sources of information, in relation to original and subsequent characteristics of the cultural heritage, and their meaning, is a requisite basis for assessing all aspects of authenticity... Aspects of the sources may include form and design, materials and substance, use and function, traditions and techniques, location and setting, and spirit and feeling, and other internal and external factors. The use of these sources permits elaboration of the specific artistic, historic, social, and scientific dimensions of the cultural heritage being examined.

The target readers of the present book are students of civil engineering and architecture, as well as any other professionals in conservation, but also cultural heritage enthusiasts with limited experience in conservation. The book is articulated into six chapters, each of them self-explanatory to allow independent reading, and the topics are detailed as follows.

Chapter 1 describes the modern understanding of conservation, focusing on the values and definitions of the main terms used throughout the text. Importance is given to the meaning of the concept of heritage structure, how it is defined and how to conserve it. In this regard, in the framework of analysis and conservation of cultural heritage structures, the basic concepts of conservation (principles) and the rules and methodology to be followed by practitioners (guidelines) are described.

Chapter 2 deals with the evolution of conservation along the past centuries, eventually leading to the present understanding. The main developments are described in chronological order (when possible), how they evolved, who were the most prominent figures and in which way they influenced the international debate. For the sake of clarity, many examples are cited accompanied by pictures.

Chapters 3 and 4 are dedicated to the history of construction. While the former is about historical materials and related structural elements, the latter illustrates the main structural systems adopted along the history. In particular, representing one the most fascinating building topologies of the past, masonry vaults are discussed extensively.

Once structural elements and systems are described, Chapter 5 gives insights on the sizing rules adopted by masons of the past. Although rudimental when compared with modern analysis tools, this ancient knowledge can give important information about the structural decisions made during the construction process. Moreover, as validated by the very existence of the building, ancient knowledge is a natural element of comparison with the first scientific theories of the 18th century. Those theories are in line with the modern understanding of limit analysis, being its theoretical and practical aspects briefly discussed.

Finally, as masonry construction is considered one of the most spread structural typologies for historical buildings, the damage and the collapse mechanisms for these constructions are described in Chapter 6. The causes of damages are differentiated according to their nature, e.g. excess of permanent loads and settlements vs. environmental and anthropogenic actions, such as earthquake or incorrect interventions, respectively.

Foreword

This book is inspired by the first two authors, lecturing activity in the International Masters in Structural Analysis of Monuments and Historical Constructions, *SAHC* (www.msc-sahc.org). SAHC was initiated in 2007, graduating since then almost 400 students from 70 countries across the world. This is a 1-year master programme jointly organized by partner universities from four countries: University of Minho (Portugal), Czech Technical University in Prague (Czech Republic), UPC/Barcelona Tech (Spain) and University of Padua (Italy). The programme also involves the Institute of Theoretical and Applied Mechanics of the Czech Academy of Sciences.

This initiative offers an advanced education programme on the conservation of cultural heritage structures, focusing on the application of scientific principles and methodologies in analysis, innovation and practice. SAHC would not have been possible without the 10-year funding from the Education, Culture and Audiovisual Executive Agency of the European Union.

The programme was the recipient of a Europa Nostra Award in 2017, category 'Education, Training & Awareness-raising', which is Europe's most prestigious cultural heritage prize. Aside from its evident significance on an international level, SAHC is impressive in its global reach. The international diversity of the students ensures that the knowledge gained has a far-reaching effect with the awareness of culture and the expertise required to protect the built cultural heritage spreading far beyond the environs of these four universities. This was noted by the Europa Nostra jury who stated: 'This project has great international value and its global outreach is a noteworthy model for other similar initiatives. The programme allows for students to view structural systems in different cultural contexts, encouraging the students to create and to develop their expertise with an increasingly valuable international perspective'.

The programme has a multidisciplinary approach comprising elements of civil engineering, conservation architecture and material science, among others. SAHC, which is based on the scientific principles of engineering, conservation and architecture, allows for a better understanding of construction systems, which in turn helps to prolong their existence.

SAHC is the only international programme that specifically addresses the conservation of historical structures. It effectively creates those professionals who have the ability to protect our shared heritage from the various threats it currently faces, such as natural decay, human interventions, climatic changes and natural hazards. The multidisciplinary aspect of this project is responding to present economic and societal needs. The programme creates the specialized expertise necessary to advance the protection of our built cultural heritage, a niche area which is becoming progressively more important.

The alumni of SAHC are now working across the world, creating a unique network of knowledge and friendship. As the course coordinator, I am indebted to our amazing students and all colleagues involved in making the course so successful. The authors acknowledge the collaboration of many colleagues contributing to the original development of the contents addressed in this book, in particular, Jiří Bláha, Jorge Branco, Ricardo Brites, Francisco Fernandes, José Luis González, Nuno Mendes, Climent Molins, Luca Pelà and Isabel Valente.

December 2018, Guimarães, Portugal
Paulo B. Lourenço, Professor, SAHC Course Coordinator

Chapter 1

Modern understanding of conservation and of heritage structures

Among the many reasons, the study of historical buildings causes an everlasting sense of wonder due to the experience of being united with the outstanding achievements of the past, feeling and learning to appreciate ancient accomplished wisdom and pioneering technology. Due to the high value attributed to historical buildings today, our modern societies are dedicating a significant effort to their interpretation and conservation. In modern times, cultural heritage has received a strong scientific attention, and today, the approach to conservation is outlined by international charters and recommendations. After centuries of debates and still ongoing discussions, cultural heritage is universally recognized as a global wealth, whose conservation preserves local and universal cultural values and contributes to human development. In this regard, the respect for authenticity represents not only one of the highest goals of conservation, but it is itself the clue to understand conservation aims, purposes, appropriate methods and expected results. With this goal, the present chapter introduces the main aspects related to the modern understanding of conservation, starting from the definition of cultural heritage buildings and ending in heritage management.

As part of the cultural heritage, this chapter is also devoted to the modern approach and conservation criteria of historical structures. The need of preserving the structure and its static model are concepts relatively recent in the history of conservation, addressed to a great extent after the Second World War. An important breakthrough was the symposium held in Ravello, Italy, in 1995, whose outcomes are the basis of the *ISCARSAH recommendations for the analysis, conservation and structural restoration of architectural heritage*. ISCARSAH is the International Scientific Committee on the Analysis and Restoration of Structures of Architectural Heritage, and it was founded by ICOMOS (International Council on Monuments and Sites) in 1996 as a forum and network of engineers and architects involved in the conservation and care of the built cultural heritage. The last part of this chapter is devoted to the description of ISCARSAH Principles and Guidelines.

Finally, although punctual references are reported within the text, for further details, the reader is referred to the original charters, available online, for example www.icomos.org or whc.unesco.org. For a glossary of terms inherent to history and culture, the reader can visit www.pc.gc.ca/en/culture/dfhd/glossaire-glossary.

1.1 DEFINITION OF BUILT CULTURAL HERITAGE AND CONSERVATION

According to the United Nations Educational, Scientific and Cultural Organization (UNESCO, 1972), 'cultural heritage' is defined as follows:

- **monuments**: architectural works, works of monumental sculpture and painting, elements or structures of an archaeological nature, inscriptions, cave dwellings and combinations of features, which are of outstanding universal value from the point of view of history, art or science (Figure 1.1a);
- **groups of buildings**: groups of separate or connected buildings which, because of their architecture, their homogeneity or their place in the landscape, are of outstanding universal value from the point of view of history, art or science (Figure 1.1b);
- **sites**: works of man or the combined works of nature and man, and areas including archaeological sites which are of outstanding universal value from the historical, aesthetic, ethnological or anthropological point of view (Figure 1.1c).

These three groups are often referred to as architectural heritage, even if the designation **built cultural heritage** (sometimes **cultural heritage buildings** is also used) is more neutral and should be preferred. As it is possible to notice, according to the Venice Charter (ICOMOS, 1964), the definition of historic monument embraces not only the single built heritage work but also the urban or rural setting in which the evidence of a particular civilization, a significant development or a historic event is found. This applies not only to great works of art but also to more modest works of the past which have acquired cultural significance with the passage of time.

According to previous descriptions, it is also possible to enumerate other subclasses of built cultural heritage, namely, **modern heritage** and **industrial heritage**. The former concerns the heritage of architecture, town planning and landscape design of the modern era, that is, 19th and 20th centuries; the latter refers to the physical remains of the history of technology and industry, such as manufacturing and mining sites, as well as power and transportation infrastructure. Two examples are shown in Figure 1.2.

Figure 1.1 Examples of built cultural heritage. (a) Religious architecture, (b) historical urban texture in ancient city centres and (c) archaeological architectural remains.

(a)   (b)

Figure 1.2 Examples of modern heritage and industrial heritage. (a) Construction in masonry, steel and concrete, and (b) masonry chimney.

Regarding **conservation**, a general and structure-oriented definition is according to

- Nara Charter, all efforts designed to understand cultural heritage, know its history and meaning, ensure its material safeguard and, as required, its presentation, restoration and enhancement (ICOMOS, 1994);
- International Organization for Standardization, all actions or processes that are aimed at safeguarding the character-defining elements of a cultural resource so as to retain its heritage value and extend its physical life (ISO 13822:2010, Annex I on Heritage Structures.

For the sake of clarity, two other definitions are given:

- **Preservation:** action or process of protecting, maintaining and/or stabilizing the existing materials, form and integrity of a cultural resource or of an individual component, while protecting its heritage value. Note that preservation is much used in American English, while conservation is much used in European English for the same concept. In Europe, preservation is used more in the context of materials or conservators (such as stone, paper, textiles, paintings, sculptures and alike), whereas conservation is almost exclusively used for the built cultural heritage;

- **Restoration**: action or process of accurately revealing, recovering or representing the state of a cultural resource or of an individual component, as it appeared in a particular period of its history, while protecting its heritage value.

Note that restoration of a cultural heritage building is a controversial concept. It encompasses, in fact, many different interpretations, ranging between reconstruction and full 're-establishment' (i.e. full recovery of an ancient building to its highest splendour, even involving the reconstruction of parts historically collapsed or possibly never built), to minimal intervention oriented to strict conservation. In this regard, it is worth noticing that a certain understanding of restoration (especially those connected to reconstruction and re-establishment) are out-of-fashion and in contradiction with modern conservation principles. Even if the word has a common root in Latin (*restauro* or to 'restore, rebuild, re-establish, renew'), its adaptation to different languages entails also different perceptions. As an example, *restauro* in Italian, per the Italian Dictionary from *Corriere della Sera*, one of Italy's oldest newspapers, reads '*Operazione e procedimento tecnico che ha lo scopo di riportare in uno stato di buona conservazione e leggibilità un bene culturale e artistico*' or 'Operations and technical procedure that aims to bring back a cultural and artistic asset to a state of good conservation and understanding', for which the best European English translation would be conservation.

1.2 CULTURAL VALUES AND AUTHENTICITY

In historical buildings, it is possible to identify sources of cultural heritage value in the following dimensions:

- cultural resource involving technical, artistic and spiritual merits;
- cultural landmark providing identity to cultures, world regions and towns;
- document on ancient knowledge, practices, culture, technology and history;
- live document providing outstanding cultural and technical achievements, from which societies can still learn and improve by studying and using;
- economic resource as a cultural/touristic attraction with an important capacity to generate secondary economy;
- contribution to cultural diversity and global cultural wealth;
- contribution to human development.

On this matter, Feilden (2003) proposed the list of values reported in Table 1.1 where, besides the cultural ones, emotional and use values are also emphasized.

Table 1.1 Values of the cultural property according to Feilden (2003)

Emotional values	Cultural values	Use values
Wonder	Documentary	Functional
Identity	Historic	Economic (including tourism)
Continuity	Archaeological and age	Social (identity and continuity)
Respect and veneration	Aesthetic and architectural	Educational
Symbolic and spiritual	Townscape	Political
	Landscape and ecological	
	Technological and scientific	

When addressing the values listed earlier, authenticity and cultural diversity play a capital role in conservation of cultural heritage buildings. According to the *Oxford English Dictionary*, authenticity is defined as 'the fact or quality of having the stated or reputed origin, provenance, or creator; not a fake or forgery'. Even if authenticity might seem related to the original creative source, it represents a subjective concept. For instance, it can be related to historical continuity of the heritage resource, including interventions in different periods of time, and the way that these have been integrated in the context of the whole (Jokilehto, 2018).

On the basis of article 9 of the Venice Charter (ICOMOS, 1964), a broader understanding of cultural diversity is clarified in the Nara document (ICOMOS, 1994). For the sake of clarity, a few points (5–8, 11–13) are reported next:

5. The diversity of cultures and heritage in our world is an irreplaceable source of spiritual and intellectual richness for all humankind. The protection and enhancement of cultural and heritage diversity in our world should be actively promoted as an essential aspect of human development.
6. Cultural heritage diversity exists in time and space, and demands respect for other cultures and all aspects of their belief systems. In cases where cultural values appear to be in conflict, respect for cultural diversity demands acknowledgement of the legitimacy of the cultural values of all parties.
7. All cultures and societies are rooted in the particular forms and means of tangible and intangible expression which constitute their heritage, and these should be respected.
8. It is important to underline a fundamental principle of UNESCO, to the effect that **the cultural heritage of each is the cultural heritage of all**. Responsibility for cultural heritage and the management of it belongs, in the first place, to the cultural community that has generated it, and subsequently to that which cares for it. However, in addition to these responsibilities, adherence to the international charters

and conventions developed for conservation of cultural heritage also obliges consideration of the principles and responsibilities flowing from them. Balancing their own requirements with those of other cultural communities is, for each community, highly desirable, provided achieving this balance does not undermine their fundamental cultural values.

11. All judgements about values attributed to cultural properties as well as the credibility of related information sources may differ from culture to culture, and even within the same culture. **It is thus not possible to base judgements of values and authenticity within fixed criteria.** On the contrary, the respect due to all cultures requires that heritage properties must be considered and judged within the cultural contexts to which they belong.

12. Therefore, it is of the highest importance and urgency that, within each culture, recognition be accorded to the specific nature of its heritage values and the credibility and truthfulness of related information sources.

13. Depending on the nature of the cultural heritage, its cultural context, and its evolution through time, **authenticity judgements may be linked to the worth of a great variety of sources of information.** Aspects of the sources may include form and design, materials and substance, use and function, traditions and techniques, location and setting, and spirit and feeling, and other internal and external factors. The use of these sources permits elaboration of the specific artistic, historic, social, and scientific dimensions of the cultural heritage being examined.

As an example, Figure 1.3 illustrates a local manufacturing technique and the work of local artisans. In the perspective of a broader meaning of authenticity, conservation projects should consider also intangible components.

(a)   (b)

Figure 1.3 Examples of authenticity. (a) Local manufacturing technique and (b) local artisan at work.

1.3 GLOBAL HERITAGE MANAGEMENT

Heritage management involves investigation, documentation, interpretation, presentation, maintenance, intervention works and education. Art, culture, education, town planning, nature and environment represent the typical background of any activity inherent to conservation.

The main targets of heritage management are local and global groups, such as inhabitants, guests, users, tourists, local community, experts and authorities. In this regard, according to the fundamental principle of UNESCO ('*The cultural heritage of each is the cultural heritage of all*'), global cultural heritage belongs to humankind, and heritage management cannot be based on other concepts than that of a common and shared valuable cultural resource.

The following subsections refer to the notions addressed in the Charter on the Interpretation and Presentation of Cultural Heritage Sites (ICOMOS, 2008).

1.3.1 Documentation

Gathering information about the artefact represents the first step of any conservation project. The sources of this information are '*all material, written, oral and figurative sources which make it possible to know the nature, specifications, meaning and history of the cultural heritage*' (ICOMOS, 1994). In this regard, Figure 1.4 provides two examples of non-conventional sources of information, namely, a drawing and incisions on the building itself. This information should be recorded and included in a dossier, together with analytical and critical reports, drawings, photos and other sources, illustrating all the stages of the conservation project, that is, before, during and after intervention. When dealing with historical heritage, besides the importance (in terms of cultural value) of recording any interaction with the object, a detailed report is fundamental because it provides the foundation for future possible interventions.

As far as built cultural heritage is concerned, documentation involves history, geometry, materials, construction techniques, morphology, structural arrangement, structural work, etc. In many cases, documentation might require sophisticated and up-to-date technologies. For instance, geometrical survey is currently often performed using drones and laser scanners, while inspection and material characterization are carried out, to the possible extent, by means of non-destructive technologies.

1.3.2 Interpretation and presentation

Interpretation refers to the full range of potential activities intended to enhance the understanding of cultural heritage and increase public awareness. This phase is based on the information gathered during documentation and represents an essential phase for an adequate identification of salient features and for the evaluation of built cultural heritage.

(a)

(b)

Figure 1.4 Examples of non-conventional sources. (a) Incision of a building elevation and (b) Colegio de Nuestra Señora de la Antigua de Monforte de Lemos (Guerra Pestonit, 2012): incised drawing on the granite floor and drawing executed on a wall with red pigment.

On the other hand, presentation regards the awareness of the public. This fundamental dimension of heritage aims to develop a greater understanding of the cultural values that monument and the surrounding site represent, as well as the respect for the role they play in contemporary society. It is noted that conservation of cultural heritage buildings is also an economical issue. As an example, tourism accounts for 10% of the world Gross Domestic Product and employment (UNWTO, 2017), and the number of tourists is expected to double in two decades. The built heritage, namely, monuments or historical centres, is a main attractor for tourism. Therefore, the need for their conservation is unquestionable, requiring societal engagement.

Interpretation and presentation are based on the following principles:

1. **Access and understanding**: interpretation and presentation pro-grammes should facilitate physical and intellectual access by the public to cultural heritage sites. This aspect may foster public awareness and engagement in the need for their protection and conservation.
2. **Information sources**: interpretation and presentation should be based on evidences gathered through accepted scientific and scholarly meth-ods as well as from living cultural traditions.
3. **Attention to setting and context**: interpretation and presentation should relate to their wider social, cultural, historical, and natural contexts and settings.
4. **Preservation of authenticity**: interpretation and presentation must respect the basic tenets of authenticity in the spirit of the Nara Document (ICOMOS, 1994), protecting the cultural values from the adverse impact of intrusive interpretive infrastructure and visitor pressure, as well as inaccurate or inappropriate interpretation.
5. **Planning for sustainability**: the interpretation plan must be sensitive to its natural and cultural environment, with social, financial and environmental sustainability among its central goals.
6. **Concern for inclusiveness**: interpretation and presentation must be the result of meaningful collaboration between heritage profession-als, host and associated communities, and other stakeholders.
7. **Importance of research, training and evaluation**: continuing research, training and evaluation are essential components of the interpretation of a cultural heritage site.

Moreover, presentation refers to the carefully planned communication of interpretive content through the arrangement of information, physical access and infrastructure at a cultural heritage site. It can be conveyed through a variety of technical means, including, yet not requiring, ele-ments such as informational panels, museum-type displays, formalized walking tours, lectures and guided tours, and multimedia applications and websites.

1.4 HERITAGE STRUCTURES AND CONSERVATION CRITERIA

A heritage structure is an existing structure or a structural component of a resource that has been recognized for its heritage value. Heritage value refers to aesthetic, historic, scientific, cultural, social or spiritual impor-tance or significance for past, present and future generations.

Together with the values already mentioned for built cultural heritage, several character-defining elements can be stated for heritage structures, for example historical materials and fabric, forms, location, spatial configuration, morphology, concept and structural design. Significant subsequent historic changes and alterations, imperfections and damage may be identified as character-defining elements too and must be respected (provided that they do not compromise the safety requirements).

On the other hand, intangible values, such as ancient culture, construction knowledge, technologies and skills, might also be identified as character-defining elements worth to be preserved (Figure 1.5). In this regard, it is worth noticing, as expressed in ISO 13822:2010 (Annex I on Heritage Structures), that judgements about heritage value may differ from culture to culture. For instance, in some geographical areas, keeping alive the traditional construction practice is privileged over the conservation of original materials. In general, authenticity for heritage structures entails several aspects. Some examples indicated by ISCARSAH recommendations (ICOMOS/ISCARSAH, 2003) are given in Table 1.2.

1.4.1 Main types of intervention on heritage structures

Important definitions, which are at the core of actions in heritage structures, are introduced in Table 1.3.

Figure 1.5 Drawing representing ancient masons and the design process.

Table 1.2 Main aspects of authenticity for heritage structures

Respect of original materials, morphology and structural arrangement

Modern technology makes possible to separate a new content from the original substance to reduce damage caused by the changing of microclimatic conditions, while providing adequate comfort for users

The Medieval Butcher's Market in Gent, Belgium

Respect for distinguishing qualities of structure and environment deriving from original form

Tübingen, Germany

Respect for original concept, materials and construction techniques
Rebuilding of Frauenkirche in Dresden, Germany

| 1860-1890 | 1966 | 2005 |

Respect for significant subsequent (historical) changes

Saint Vitus in Prague, Czech Republic:
Baroque roof
Renaissance gallery
Gothic cathedral

(*Continued*)

Table 1.2 (Continued) Main aspects of authenticity for heritage structures

Respect for alterations or imperfections (deformations) that have become part of the history of the structure, provided that they do not compromise the safety requirements

Leaning Tower in Pisa, Italy
Crooked Spire in Chesterfield, England

Table 1.3 Definition of main structural interventions

Stabilization
Action aimed at stopping a deteriorating process involving structural damage or material decay. Stabilization is also applied to actions meant to prevent the partial or total collapse of a deteriorated structure. Stabilization is often applied to archaeological remains or structures having partially collapsed in historical times. It is also applicable to constructions suffering chemical or physical attacks causing a gradual decay of materials or structures having experienced significant destruction due to an extraordinary action such as earthquake

Emergency stabilization (or emergency action)
Action or process implemented urgently to secure a structure that has insufficient safety level for a limited time until conservation works can be implemented

Repair
Action to recover the initial mechanical or strength properties of a material, structural component or structural system. Repair is applicable to cases where a structure has experienced a deterioration process having produced a partial loss of its initial performance level. In the context of conservation of historical structures, repair is not meant to correct any historical deterioration or transformation (including those manmade)

(Continued)

Table 1.3 (Continued) Definition of main structural interventions

that only affects the appearance or formal integrity of the building and does not compromise its stability. Repair should be only used to improve structures having experienced severe damage actually conveying a loss of structural performance and thus causing a structural insufficiency with respect to either service or ultimate loading

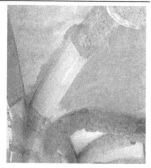

Strengthening

Action providing additional strength to the structure. Strengthening may be needed to resist new loading conditions and uses, to comply with a more demanding level of structural safety or to respond to increasing damage associated with continuous or long-term processes. Strict conservation will normally require stabilization or repair operations. Conversely, rehabilitation will frequently lead to strengthening operations

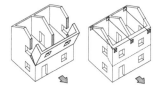

Seismic retrofit (or seismic upgrading)

Modification of structures to make them more resistant to seismic activity ground motion or soil failure due to earthquakes. Seismic retrofit can be achieved by means of appropriate strengthening. However, strengthening is not the only possibility to improve the seismic response of a structure, since seismic retrofit can also be achieved through alternative strategies

Rehabilitation – first definition

Upgrading of a structure to comply with modern uses and standards. Rehabilitation may lead in some cases to alter the structure to an extent incompatible with the conservation principles. It may constitute an activity substantially different to strict conservation

Rehabilitation – second definition

Action or process of making possible a continuing or compatible contemporary use of a cultural resource or an individual component, through repair, alterations and/or additions while protecting its heritage value

The problem with this definition is that making possible a modern use according to modern standards and codes may be incompatible with sound protection of heritage value. Rehabilitation may in some cases require significant transformation with loss of authenticity and cultural value

(Continued)

Table 1.3 (Continued) Definition of main structural interventions

According to these definitions, rehabilitation may
constitute an activity substantially different from
strict conservation and may possibly induce
significant alteration to the structure to an extent
incompatible with the restoration principles

Minimal intervention

Intervention that balances safety requirements with
the protection of character-defining elements,
ensuring the least harm to heritage values – ISO
13822:2010 (Annex I on Heritage Structures)

Non-destructive testing

Experimental determination of the mechanical,
physical or chemical properties of materials or
structural members that does not cause any loss of,
or damage to, the historic fabric. Sometimes, the
term **non-invasive technique** is used, but this can
be used also for interventions and not only for
diagnostic tools

Sonic testing

Minor destructive testing

Experimental determination of the mechanical,
physical or chemical properties of materials or
structural members that cause minimal and easily
repairable damage to the historic fabric

Micro-drilling

Incremental approach

A step-by-step procedure in which the behaviour of
the structure is monitored at each stage, and the
data acquired is then used to provide the basis for
further action – ISO 13822:2010 (Annex I on
Heritage Structures)

Crack monitoring using a tell-tale

1.4.2 Modern approach to heritage structures conservation

In essence, it is possible to distinguish between two different historical attitudes towards heritage structures (and the way they can be studied and preserved), namely, past and modern understanding (for further details, the reader is referred to Chapter 2 where these concepts are refined and discussed within the history of conservation). In order to show the main differences, Table 1.4 depicts past and modern approaches for conservation in case of a damaged arched structure. In the example considered, the thrust of the arch induces an outward movement of the springings, with cracks in the key of the arch and at the arch supports, and out-of-plumb movement of the walls to the outside.

As it is possible to notice from Table 1.4, the past understanding generally leads to invasive strengthening measures that deeply alter the structure. Many past interventions were due to several incorrect assumptions, such as

Table 1.4 Differences between past and modern approach to heritage structures

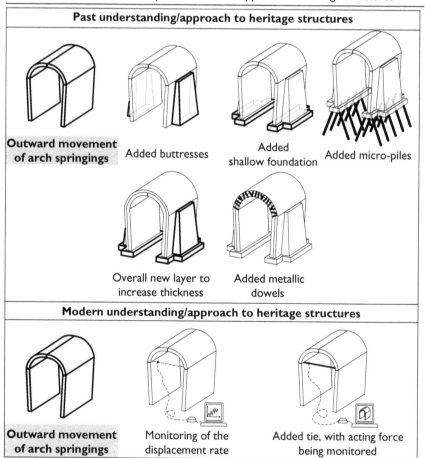

Past understanding/approach to heritage structures		
Outward movement of arch springings	Added buttresses	Added shallow foundation	Added micro-piles
	Overall new layer to increase thickness	Added metallic dowels	

Modern understanding/approach to heritage structures		
Outward movement of arch springings	Monitoring of the displacement rate	Added tie, with acting force being monitored

(1) mistrust (due to ignorance) towards original or ancient materials and original resisting resources, (2) a blind confidence in modern materials and technologies, (3) a lack of recognition on the value of original/ancient structure and original structural features, and (4) a lack of recognition on the importance of studies previous to any intervention. As a result, significant negative experience was accumulated, requiring not only subsequent repair and replacement, that is, the so-called *restoration of restorations*, but also leading to important fabric losses, affecting the significance and requiring unnecessary financial resources.

This approach was partially (and erroneously) fostered by the Athens Charter (1931) that recommended the use of concrete and other modern materials and techniques (see Section 2.3). In addition to this, still against modern understanding, the Athens Charter recommended to hide new materials and components to avoid altering the historical aspect of the building.

Quite in opposition, modern understanding is characterized by a deep knowledge and respect towards authenticity of structure and structural principles governing its response. Acknowledging the importance of previous studies (comprising historical, material and structural aspects), the real causes of possible damage or alterations must be investigated to propose respectful interventions. These should be minimal, non-intrusive and reversible.

This is the line promoted by the Venice Charter (ICOMOS, 1964), where the use of traditional or historical materials for stabilization or restoration is implicitly recommended. The use of modern materials and techniques is limited to the cases where it is not possible to adopt traditional and historical techniques for the sought purpose. In any case, it must be possible to distinguish new materials or components from the original ones. According to the more recent ISCARSAH recommendations (ICOMOS/ISCARSAH, 2003), the basis for the design of new interventions are safety evaluation (and requirements), compatibility (between original and newly added materials/components), least invasiveness, non-obtrusiveness, durability, reversibility and controllability (monitoring).

Noticing the complexity of process, it is generally accepted (ICOMOS/ISCARSAH, 2003) that the study of historical structures should not only be based on calculations but also integrate a variety of complementary activities. Structural analysis of historical structures constitutes, in fact, a multidisciplinary, multifaceted activity that requires a clever integration of different approaches and sources of evidence (Roca et al., 2010). In this regard, it must be acknowledged that conventional calculation techniques and legal codes or standards oriented to the design of modern constructions may be difficult to apply or are even inapplicable to ancient structures.

1.5 ICOMOS/ISCARSAH RECOMMENDATIONS

The ISCARSAH *recommendations for the analysis, conservation and structural restoration of architectural heritage* include two parts, namely,

Principles (basic concepts of conservation) and **Guidelines** (the rules and methodology to be followed by practitioners). Next, the Principles are addressed with clear reference to the articles of the original document, whereas only a synthetic description is provided for the Guidelines, referring the reader to the original document for further details.

1.5.1 Principles

As stressed in the purpose of the document, only the Principles have the status of an approved/ratified ICOMOS document. The articles inherent to the **general criteria, research and diagnosis,** and **remedial measures and controls** are reported in Tables 1.5–1.7, respectively.

Table 1.5 Principles from ISCARSAH recommendations: general criteria

1.1	Conservation, reinforcement and restoration of architectural heritage requires a **multidisciplinary approach** 
1.2	**Value and authenticity** of architectural heritage **cannot be based on fixed criteria,** because the respect due to all cultures also requires that its physical heritage be considered **within the cultural context** to which it belongs 
1.3	The value of architectural heritage is not only in its appearance but also in the **integrity of all its components** as a unique product of the specific building technology of its time. In particular, the removal of the inner structures maintaining only the façades does not fit the conservation criteria 

(Continued)

Table 1.5 (Continued) Principles from ISCARSAH recommendations: general criteria

1.4	When any **change of use or function** is proposed, all the conservation requirements and safety conditions have to be carefully taken into account	
1.5	**Restoration of the structure** in architecture heritage is not an end in itself but a means to an end, which is the **building as a whole** **Note**: *A global understanding of the building is needed*	
1.6	The peculiarity of heritage structures, with their complex history, requires the organization of studies and proposals in precise steps that are similar to those used in medicine. **Anamnesis, diagnosis, therapy and controls** corresponding, respectively, to the searches for significant data and information, individuation of the causes of damage and decay, choice of the remedial measures and control of the efficiency of interventions. In order to achieve cost effectiveness and minimal impact on architectural heritage using funds available in a rational way, it is usually necessary that the study repeats these steps in an **iterative process**	
1.7	**No action should be undertaken without having ascertained the achievable benefit** and harm to the architectural heritage, **except in cases where urgent safeguard measures are necessary** to avoid the imminent collapse of structures (e.g. after seismic damages); those urgent measures, however, should **when possible avoid modifying the fabric in an irreversible way**	

Table 1.6 Principles from ISCARSAH recommendations: research and diagnosis

2.1	Usually a **multidisciplinary team**, to be determined in relation to the type and the scale of the problem, should work together from the first steps of a study – as in the initial survey of the site and the preparation of the investigation programme 
2.2	Data and information should first be processed approximately, to establish a **more comprehensive plan of activities in proportion to the real problems** of structures 
2.3	A **full understanding of the structural and material characteristics** is required in conservation practice. Information is essential on the structure **in its original and earlier states**, on the techniques that were used in the construction, on the alterations and their effects, on the phenomena that have occurred and, finally, **on its present state** 
2.4	**In archaeological sites** specific problems may be posed because **structures have to be stabilized during excavation** when knowledge is not yet complete. The structural responses to a 'rediscovered' building may be completely different from those to an 'exposed' building. Urgent site-structural solutions, required to stabilize the structure as it is being excavated, should not compromise the complete building's concept form and use 

(*Continued*)

Table 1.6 (Continued) Principles from ISCARSAH recommendations: research and diagnosis

2.5	**Diagnosis is based on historical, qualitative and quantitative approaches**; the qualitative approach is mainly based on direct observation of the structural damage and material decay as well as historical and archaeological research; the quantitative approach is mainly based on material and structural tests, monitoring and structural analysis	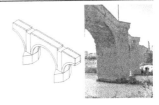
2.6	Before making a decision on structural intervention, **it is indispensable to first determine the causes of damage and decay** and then to evaluate the safety level of the structure	
2.7	**The safety evaluation**, which **is the last step in the diagnosis** where the need for treatment measures is determined, **should reconcile qualitative with quantitative analysis**, including direct observation, historical research, structural analysis and, if it is the case, experiments and tests	
2.8	Often the application of the same safety levels as in the design of new buildings requires excessive, if not impossible, measures. In these cases, **specific analyses and appropriate considerations may justify different approaches** to safety Note: *It may be possible to adopt a safety improvement approach instead of a full compliance with a code for modern structures*	
2.9	All aspects related to the acquired information, the diagnosis including the safety evaluation and the decision to intervene should be described in an '**explanatory report**'	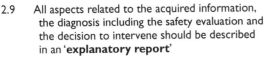

Table 1.7 Principles from ISCARSAH recommendations: measures and controls

3.1	**Therapy should address root causes rather than symptoms**	
3.2	**The best therapy is preventive maintenance** **Note:** *Adequate maintenance can limit or postpone the need for subsequent intervention*	
3.3	**Safety evaluation and an understanding of the significance of the structure** should be the basis for conservation and reinforcement measures **Note:** *Understanding of historical and cultural significance is needed (respect for authenticity)*	
3.4	**No actions should be undertaken without demonstrating that they are indispensable** **Note:** *This means that no intervention should be implemented without ascertaining the likely benefit and harm to the built cultural heritage and long-term side effects*	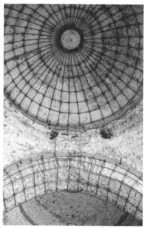

(Continued)

Table 1.7 (Continued) Principles from ISCARSAH recommendations: measures and controls

3.5	Each intervention should be in proportion to the safety objectives set, thus **keeping intervention to the minimum to guarantee safety and durability** with the least harm to heritage values **Note:** *The real need must be assessed and only indispensable actions are to be implemented*

3.6	The design of intervention should be based on a clear understanding of the kinds of **actions that were the cause of the damage and decay as well as those** that are taken into account for the analysis of the structure **after intervention,** because the design will be dependent upon them

3.7	The choice between 'traditional' and 'innovative' techniques should be weighed up on a case-by-case basis and preference given to those that are **least invasive and most compatible with heritage values**, bearing in mind safety and durability requirements

3.8	**At times,** the difficulty of evaluating the real safety levels and the possible benefits of interventions may suggest '**an observational method**', that is, an incremental approach, starting from a minimum level of intervention, with the possible subsequent adoption of a series of supplementary or corrective measures

(Continued)

Table 1.7 (*Continued*) Principles from ISCARSAH recommendations: measures and controls

3.9	Where possible, **any measures adopted should be 'reversible'** so that they can be removed and replaced with more suitable measures when new knowledge is acquired. Where they are not completely reversible, interventions should not limit further interventions	
3.10	The characteristics of materials used in restoration work (in particular new materials) and their **compatibility with existing materials** should be fully established. This must include long-term impacts, so that undesirable side effects are avoided **Note:** *It is clear that compatibility (chemical, mechanical, rheological, thermal, physical, etc.) is a necessary condition but not sufficient to accept a product, because its benefit has to be demonstrated*	
3.11	**The distinguishing qualities** of the structure and its environment, in their original or earlier states, **should not be destroyed**	
3.12	Each intervention should, as far as possible, **respect the concept, techniques and historical value** of the structure and the historical evidence that it provides	
3.13	**Intervention should be the result of an overall integrated plan** that gives due weight to the different aspects of architecture, structure, installations and functionality **Note:** *Symbolic, historical and documental aspects need to be considered*	

(*Continued*)

Table 1.7 (Continued) Principles from ISCARSAH recommendations: measures and controls

3.14	**The removal or alteration** of any historic material or distinctive architectural features **should be avoided whenever possible** 
3.15	**Deteriorated structures whenever possible should be repaired rather than replaced** 
3.16	**Imperfections and alterations**, when they have become part of the history of the structure, should be maintained, provided that they do not compromise the safety requirements 
3.17	**Dismantling and reassembly** should only be undertaken as an optional measure required by the very nature of the materials and structure when conservation by other means impossible, or harmful 

(Continued)

Table 1.7 (Continued) Principles from ISCARSAH recommendations: measures and controls

3.18	**Provisional safeguard systems** used during the intervention should show their purpose and function without creating any harm to heritage values

3.19	Any proposal for intervention must be accompanied by a **programme of control** to be carried out, as far as possible, **while the work is in progress**
3.20	**Measures that are impossible to control during execution should not be allowed**

3.21	**Checks and monitoring during and after the intervention** should be carried out to ascertain the efficacy of the results

3.22	**All the activities of checking and monitoring should be documented** and kept as part of the history of the structure

1.5.2 Guidelines

1.5.2.1 General aspects

A combination of scientific and cultural insight, together with experience, is indispensable for the study of built cultural heritage. As stressed in the Principles, a scientific and multidisciplinary approach involving historical investigation, inspection, monitoring, structural modelling and analysis is requested. For instance, historical research can discover phenomena involving structural behaviour, while historical questions may be answered by considering the structural behaviour.

In this rather broad scenario, structural engineering represents the scientific support necessary to safeguard the cultural and historical value of the building. With this purpose, the practitioner should adopt the so-called *holistic approach*, according to which the building is and works as a spatial environmental system, and it must be studied as such. This means that the significance of any element, object and feature should be analysed within the context of the entire structure. In other words, the construction must be understood as a whole, rather than as an assemblage of individual elements.

Any planning for structural conservation requires both **qualitative data** (direct observation of material decay and structural damage, historical research, etc.) and **quantitative data** (based on specific tests and mathematical models of the kind used in modern engineering). If on the one hand, this strategy is hardly contemplated in conventional codes and methods, on the other hand, the integration of qualitative and quantitative evidences is particularly aimed to overcome their limitations. Codes prepared for the design of modern structures, in fact, are often not applicable to historic structures. For example, the enforcement of seismic and geotechnical codes can lead to a severe underestimation of real structural behaviour and/or drastic, often unnecessary, strengthening measures.

However, given the subjectivity of the aspects involved in the study, namely, the uncertainties in the assumed data and the difficulties of a precise evaluation of the involved phenomena, the conclusions may be of uncertain reliability. Accordingly, it is important to clearly show these aspects in an **explanatory report** that contains a careful and critical analysis of the safety of the structure.

From the operative point of view, the recommendations propose an approach based on the following phases: (1) **diagnosis** (identify causes of damage and decay), (2) **safety evaluation** (determine acceptability of safety levels by analysing the present condition of structure and materials), and (3) **design of remedial measures** (layout repair or strengthening actions to ascertain the required safety). The approach is illustrated in Figure 1.6 by means of a case study (Church of Saint Torcato in Guimarães, Portugal). In particular, diagnosis and safety evaluation of the structure are two consecutive and related stages on the basis of which the effective need for and the extent of treatment measures are determined. If these stages are performed incorrectly, the resulting decisions will be arbitrary: poor judgement may result in either too conservative, and therefore, heavy-handed conservation measures or inadequate safety levels. Too conservative actions may have a significant cost in terms of loss of cultural value and authenticity, involving important alterations or transformations of the structure.

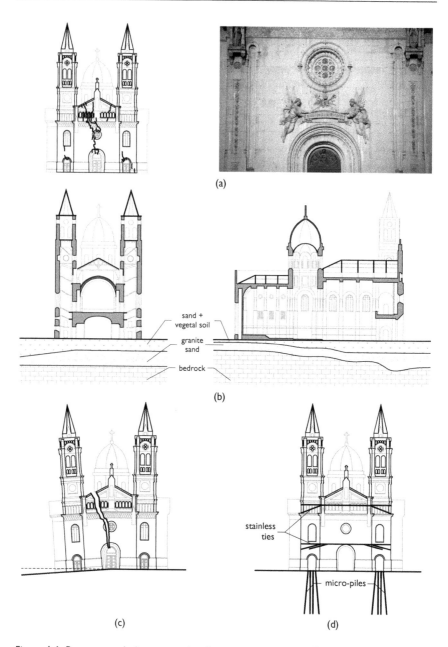

Figure 1.6 Recommended approach for conservation of heritage structures: Church of Saint Torcato in Guimarães (Portugal) as an illustrative example. (a) **Damage survey** and detail of the façade above the main door, (b) **diagnosis** and differential settlements due to non-homogenous soil stratification, (c) **safety evaluation** and (d) **remedial measures**.

1.5.2.2 Diagnosis

According to the Principles, diagnosis is supported by qualitative and quantitative data, that is, historical investigation and inspection of the present conditions on one side, and monitoring and structural analysis on the other side. Based on the information gathered through these approaches, it may be possible to simulate the effect of historical actions (including gravity, settlements, and earthquake) and, therefore, investigate the possible influence of such actions on the present condition of the building.

As far as the historical survey is concerned, its purpose is to understand the conception and significance of the building, the techniques and the skills used in its construction, the subsequent changes in both the structure and its environment, and any events that may have caused damage (e.g. wars, earthquakes and floods). In short, any damage, failure, reconstruction, addition, change, restoration work, structural modification and change of use that have led to the present conditions should be noted and interpreted. As an example, according to historical research, Roca et al. (2013) investigated the possible influence of the long and gradual construction process as well as of later long-term deformation on the final condition of the Mallorca Cathedral (Figure 1.7).

The main concern of historical investigations is related to the fact that historical documents were usually prepared for purposes other than technical issues, mostly related to accounting information (purchase of materials, payment of salaries, etc.). In this regard, a sound cooperation between historians and structural analysts is needed so that the first can infer

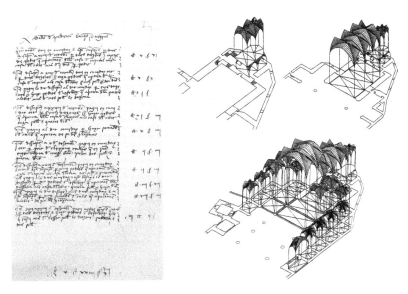

Figure 1.7 Investigation of the construction process of Mallorca Cathedral from medieval manuscripts.

historical relevant facts and succeed in identifying structurally meaningful information.

Regarding inspections, they can be considered according to two different standpoints: (1) survey of the structure (by direct observation) and (2) on-site research and laboratory testing. The direct observation is aimed to identify decay and damage as well as any ongoing environmental effects on the building. Furthermore, the survey can help in deciding whether there is an immediate danger that requests urgent measures to be undertaken. The on-site research regards internal morphology, properties of the materials (mechanical, physical, chemical), evaluation of stresses and deformation of the structure, and the presence of any discontinuities within the structure. As a general rule, non-destructive tests should be adopted. An incremental approach can be used for this purpose. If tests are necessary, cost–benefit analysis should be carried out (benefit of information and possibility of optimized structural intervention against the loss of culturally significant material). When possible, different methods should be used and the results compared.

Another source of structural information is monitoring, which can provide insight into the condition of the structure and the presence of active processes associated to incremental damage (as an extreme case, monitoring can also trigger an alarm). The results of a monitoring programme should be always analysed in the light of the historical nature of construction. Considering the long-time processes, which usually encompass decades, centuries or millennia (Figure 1.8), the challenge of the analyst is to develop hypotheses or conclusions on the condition of the structure, and on the phenomena acting upon it, according to just a small time frame (Roca et al., 2008).

From the technical point of view, monitoring can be catalogued as static or dynamic: the former usually aims to record slow varying changes in deformation, cracks, temperatures, etc.; the latter is used to record fast varying data, such as accelerations, and to characterize the dynamic response of

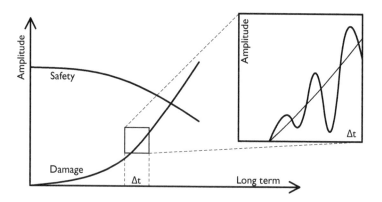

Figure 1.8 Monitoring as a short window over historical time (Roca et al., 2008).

the building. In any case, monitoring can be used to acquire useful information at different stages of the study or works on the building, such as
before the intervention (for diagnosis purposes), during the implementa-
tion of strengthening measures, or in long-term survey for monitoring the
strengthened building. Observation can also support possible auxiliary
remedial actions.

Whatever the case may be, it is worth noting that in historical buildings,
deformation and damage develop as a superimposition of different phenomena, some of which act persistently or continuously, some cyclically or
periodically, others occurring only on well-defined occasions. The analyst
should unravel the registered information into its different components,
so that they can be related to different actions or phenomena, for example
cyclic and reversible such as the seasonal effects (e.g. due to temperature,
humidity or water table level), versus long-term accumulation of irreversible components. As a general rule, the breakdown of the captured response
should consider cyclic, circumscribed and monotonic components, of which
the latter two might be associated to possible active damaging processes
(Roca et al., 2008). In this regard, monitoring outcomes can be better interpreted with the help of structural analysis.

1.5.2.3 Structural model

Another important tool currently widely adopted in the diagnosis phase is
structural analysis, which require the consideration of aspects such as

- geometry and morphology (structural form, internal composition,
  connections between the structural elements, etc.);
- material properties;
- actions (mechanical, physical, chemical, etc.);
- existing alterations and damage (cracks, constructional mistakes, dis-
  connections, crushing, leanings, etc.);
- the interaction of the structure with the soil (except in the cases where
  it is judged to be irrelevant).

The conceptual process that leads the analyst to understand, define, quan-
tify, visualize and simulate the effects of these quantities goes under the
name of **modelling**. Being a representation of the reality, a model is necessarily an approximation and represents a limited number of the (assumed)
most influencing features, reaching a compromise between realism and
cost. Given the difficulties of this process, which implicitly involves concepts and hypotheses (e.g. on the description of the actions, or mechanical,
geometrical and morphological properties), no model can describe the reality exactly, and the analyst is requested to calibrate and validate it.

Consequently, based on observations and empirical or experimental information, an iterative procedure is normally adopted to validate (or refuse)

hypotheses. Only when the procedure is over, that is, when results simulate correctly the observation, the model can be used for safety predictions and evaluation of strengthening measures. For instance, Ciocci et al. (2018) used the post-earthquake damage in Ica Cathedral (Peru) to validate the numerical model (Figure 1.9) and, subsequently, to predict the capacity of the structure before and after strengthening works.

Structural modelling is currently considered as a numerical tool almost exclusively, while in the past physical and scaled models were widely used. A clear example is the model presented by Brunelleschi as a proposal for the dome of the Florence Cathedral (Figure 1.10a). In addition, models were also used in the past to solve problems of construction (e.g. stereotomy). For instance, in 2012, during the excavation to expand the Florence Cathedral Museum (an area that may have served as a workshop for the construction of the dome), a scaled model of a dome measuring around 3 m in circumference was uncovered (Figure 1.10b). Whether or not it was built by the famous designer is not clear, but the herringbone bond of the bricks (*a spina di pesce*), not common in Florence, provides a solid hint for this hypothesis. Perhaps the model was used to verify the possibility of building the dome without formwork.

Besides these practical applications, scaled models had also a structural meaning: as long as safety is based on stability (a matter of geometry and not of strength) and dynamic effects are not considered, the capacity of a scaled model coincides with the one of the full-scale construction (the reader is referred to Chapter 5 for further details).

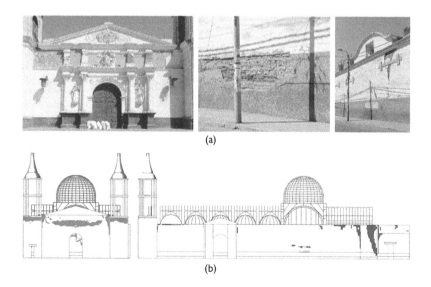

(a)

(b)

Figure 1.9 Investigation of the damage in Ica Cathedral, Peru (Ciocci et al., 2018). (a) Photos of the damage after the Pisco 2007 earthquake and (b) structural model (dark grey indicates larger cracks).

(a)   (b)

Figure 1.10 Dome of the Florence Cathedral. (a) Wooden model exposed in the
Museo dell'Opera del Duomo (Museum of the Works of the Cathedral)
in Florence and (b) the possible scale model with herringbone pattern
uncovered in 2012.

Thanks to the work of architects, engineers and researchers, the concept
of scaled models and their extension to real-scale structures have fully developed in the past centuries. The Buckingham π theorem for dimensional analysis was stated in 1914, having been originally generalized by Vaschy in 1892.

The three-dimensional model made by Antoni Gaudí for the Church of
Colònia Güell is an extraordinary example of the use of models for structural purpose in the 20th century (Figure 1.11a). The Crypt of the Church
of Colònia Güell is the only component to have been built as part of a larger
project for the church. Following several restorations, among which the
one of 1939 after the damages produced by the Civil War in 1936, a more
recent restoration was carried out, strongly criticized by Catalan intellectuals. Considering the concept of the catenary curve (see Section 5.1.3),
Gaudí built an 1:10 scale upside-down model to explore the organic design
of the church; the loads were modelled with small bags full of birdshot
hanging from the model (Figure 1.11b).

Regarding structural analysis (rather than the design) of historical constructions, Mark et al. (1973) applied the photo-elasticity technique to
experimentally determine the stress distribution of two bays of the 13th-
century choir vaults of Cologne Cathedral. A few years later, among the
study of several other Gothic cathedrals, Mark (1982) performed the plane
analysis of Mallorca Cathedral cross section. Figure 1.12 collects the
pictures inherent to the two works.

In the same period (1960s/1970s), the Finite Element Method (FEM)
became a widely disseminated numerical tool. Today, it is the most used
method of structural analysis, a sort of virtual laboratory for testing
buildings. Although originally conceived for the analysis of linear elastic

(a)   (b)

Figure 1.11 Antoni Gaudí's model of Colonia Güell. (a) Old photograph of the original
model and (b) the modern copy where the bags (originally full of birdshot)
are used to simulate the weight of the supported elements.

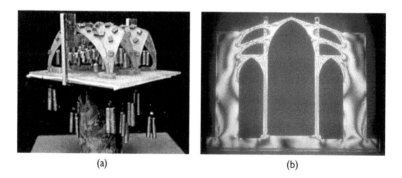

(a)   (b)

Figure 1.12 Photo-elastic analysis of Gothic structures. (a) Cross vaults of Cologne
Cathedral (Mark et al., 1973) and (b) cross section of Mallorca Cathedral
(Mark, 1982).

structures, FEM is nowadays implemented to perform complex analyses,
taking into account the nonlinear behaviour of the material (e.g. masonry
as an anisotropic composite, featuring cracking and crushing), as well as
load history, geometry, boundary conditions, settlements, morphology,
connections, previous damages and possible alterations (Lourenço, 2002;
Roca et al., 2010). Figure 1.13 shows some numerical analyses recently
performed in large monumental structures.

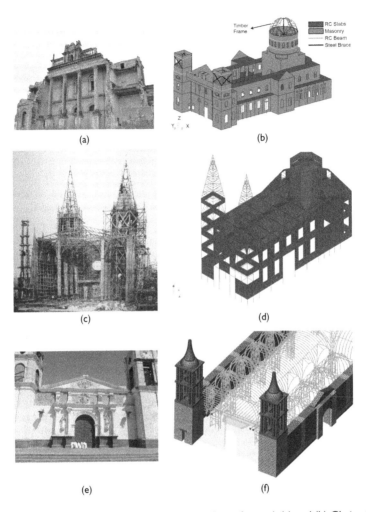

Figure 1.13 Examples of numerical analyses recently performed. (a) and (b) Christchurch cathedral, New Zealand (Silva et al., 2018); (c) and (d) San Sebastian Basilica, Manila, Philippines (O'Hearne et al., 2018); and (e) and (f) Ica Cathedral, Peru (Ciocci et al., 2018).

1.5.2.4 Safety evaluation and explanatory report

While the object of diagnosis is to identify the causes of damage and decay, safety evaluation provides the basis to understand the effective need for and the extent of treatment measures. When dealing with cultural heritage buildings, safety evaluation is aimed at finding the best compromise between the different principles described earlier (e.g. minimum intervention), while avoiding excessive risk for the human life. If diagnosis and safety evaluation are performed incorrectly, the resulting decisions will be arbitrary, either too conservative or providing inadequate safety levels. Too conservative measures may represent a loss of the cultural value of the building due to unnecessary transformations or alterations of the structure. On the other hand, insufficient measures may lead to an unacceptable risk to people and property.

Regarding legal issues, it is important to stress that codes prepared for the design of modern structures may be inappropriate for historic structures. They are based on calculation approaches which may fail to recognize the real structural behaviour and the safety condition of ancient constructions, leading to drastic and often unnecessary strengthening measures. For instance, modern codes of practice suggest the application of safety factors to take into account the various uncertainties. A conservative application of this approach may not be problematic for new structures, where safety can be increased with modest increases in member size and cost. However, in historical structures, conservative requirements to improve the strength may lead to excessive or unnecessary loss of historic fabric or changes in the original conception of the structure. A more flexible and broader approach, in which calculations are not the only source of evaluation, is requested.

The final decision regarding the safety of the structure is up to the analyst and should be based on the combination and integration of four different approaches, namely, the **historical, qualitative, analytical** and **experimental approaches,** each giving a separate contribution (Table 1.8). Given the possible subjectivity of these aspects, the uncertainties in the assumed data and the difficulties of a precise evaluation of the phenomena, the conclusions may be of uncertain reliability. It is important, therefore, to clearly demonstrate the care taken in the development of the study, as well as the reliability of data and results in an **explanatory report.** The report should also mention the hypotheses and degree of caution being exercised at each stage, the analysis of the qualitative data and the careful judgement and proposal of the 'best' intervention. A second opinion seems recommended in case of proposals for major interventions.

Table 1.8 The four different approaches supporting safety evaluation

Historical
Knowledge of what has occurred in the past can help to forecast the future behaviour and can be a useful indication of the level of safety provided by the present state of the structure. History is the most complete, life-size, experimental laboratory

Note:
1755 Lisbon earthquake print

(Continued)

Table 1.8 (Continued) The four different approaches supporting safety evaluation

Qualitative (inductive procedure)

Much is learned from the comparison between the present condition of structure and that of similar structures whose behaviour is understood or can be observed upon a specific event. Experience gained from analysing and comparing the behaviour of different structures can enhance the possibility of extrapolations and provide a basis for assessing safety

Note:

Characterizing and validating typical failure mechanisms for buildings is the support for the limit analysis using macro-elements

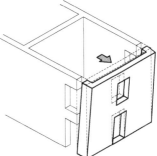

Analytical (deductive procedure)

The methods of modern structural analysis allow, on the basis of certain hypotheses (theory of elasticity, theory of plasticity, frame models, finite element modelling, etc.), to draw conclusions based on mathematical calculations

Note:

An example of modelling and analysing a structure to obtain quantitative predictions on its response subjected to different actions (red indicates larger cracks)

Experimental

Specific tests (such as proof test loading a floor, a beam, etc.) provide a measure of safety, even if load proof tests are usually applicable to single elements (or structural parts) rather than to the building as a whole. Additionally, monitoring allows to understand the structure performance under existing or additional loading

Note:

The tests can also monitor the behaviour at different stages of the conservation works (observational approach) and the acquired data can be used to provide the basis for possible further action

Chapter 2

History of conservation

Conservation is a complex concept having experienced a very diverse understanding in different historical periods and cultures. Consequently, it is not easy to discuss about the definition or goals of this activity along the history and even to clearly understand when it was born. For instance, if conservation is understood as all actions needed to maintain and use a construction for different purposes, conservation was born together with the human kind, or at least with the onset of civilization about 10,000 years ago. Differently, the modern concept of cultural heritage, regarded as an evidence of the past to be preserved and transmitted to future generations, is quite young, and has been addressed only during the 18th and 19th centuries in a fervent cultural environment (Carbonara, 1996).

With much doubt, centuries of philosophical, aesthetic and technical progress were needed to develop and articulate the modern criteria of conservation (Feilden, 2003), as discussed in Chapter 1. In the present chapter, the historical developments of this process are just outlined with a focus on more technological and structural aspects, the reader being referred to literature (Carbonara, 1996; Fernández Alba et al., 1997; Esponda, 2003; Feilden, 2003; Croci, 2008; Jokilehto, 2018) as well as the International Council on Monuments and Sites (ICOMOS) website (Charters and Recommendations) for further details.

2.1 FROM ANTIQUITY TO THE 18TH CENTURY

In ancient times, despite the singularities related to conservation (together with the circumstances and the knowledge of that time), the interventions on historical buildings were basically aimed at safeguarding the spiritual or economic value of the construction, according to the current time standpoint. One of the most ancient examples of interventions related to the preservation of the spiritual value of monuments regards the statue of Ramses II in the temple of Abu Simbel, around 1200 BC. Given the still profound respect for such influencing leader (and not for preserving any historical evidence), a masonry pier was built to sustain one of the arms (Figure 2.1).

(a)   (b)

Figure 2.1 Abu Simbel temple. (a) External view and (b) masonry pier added as a support
for one of the arms.

In turn, a historical building had an economic value that, somehow,
guaranteed the survival of portions of the construction (most of the time,
foundations or walls). The possibility of re-using existing elements meant
materials and workmanship saving. Interventions were often dramatic,
for example removal, substitution or addition of large parts. Assuming no
distinctions between the past and the present, the interventions on existing buildings and portions of them were allowed to satisfy new needs,
requirements or aesthetic values. In a sort of endless updating, the built
environment became, layer-by-layer, an integral part of the new construction (approach that goes under the name of *integration*).

An extraordinary example of integration is the Basilica of Saint Clement in
Rome, which suffered many transformations: (1) probably a republican villa
and warehouse destroyed in the Great Fire of 64 AD; (2) home of a Roman
nobleman and public building built on the existing foundations; (3) perhaps
site of clandestine Christian worship in the 1st century; (4) partly used as a
mithraeum (site for the cult of Mithras) starting from the 2nd century; (5)
first basilica, built on the Roman buildings in the 4th century; and (6) present
basilica, built on top of the first basilica, just before the year 1100. The two
basilicas and the archaeological remains of the 1st century are still visible.

In addition to this evolutionary approach, in many cases, only the location of the building was considered important, and the building parts
were demolished and substituted by new ones better adapted to the contemporary architectural styles and ideology of the corresponding societies
(so-called *appropriation*). For instance, former Paleo-Christian churches
were demolished and substituted by Romanesque ones, which were later
demolished and substituted by Gothic structures, with the relevance given
to the place rather than to the building.

During the Roman Era, to restore (in a subjective manner) the original aspect and the uniformity of the building, the most common form of conservation was *reconstruction*. The Pantheon is an extraordinary example of this approach: built in 27 BC, renovated two times in 81–117 AD, completely rebuild in 120–124, restored in 138–161 and, finally, renovated in 202. Nonetheless, with the advent of Christian religion, countless pagan temples were transformed into churches. For instance, the Pantheon became the Church of Saint Mary and the Martyrs in 608–615, fortunately without further important interventions. On the other hand, important and severe alterations and destructions were necessary to transform the Parthenon (Athens) into the church dedicated to the Virgin Mary in 662 AD.

In the Middle Ages, to preserve the economic relevance of the place, churches and basilicas were built upon the ruins of old Roman buildings, often with radical interventions, for example the construction of vaults to replace wooden ceilings and the addition of the transept as a new architectural element. Furthermore, this approach was also aimed at 'enhancing' the new construction with ancient glory, with a strong symbolic value. This is the case of Charlemagne. His willingness to renew the ancient Roman Empire included the use of ancient materials and elements from Roman Era, with the clear intention of reflecting his ideals and cultural values into the built heritage.

In this early period, although the idea of conservation was not formed yet, strengthening measures were undertaken to increase the stability of monuments as needed. One of the most interesting cases regards the conservation of Hagia Sophia, Istanbul, Turkey (originally from the 6th century AD) during the 16th century. The Ottoman architect Mimar Sinan proposed to build an impressive buttressing system at the exterior of the basilica to support the thrust of the internal vaulted structures (Figure 2.2). Also in Istanbul, an important intervention was performed in the Chora Church (6th century AD) using ties and flying arches (Figure 2.3).

It is only during the Renaissance (starting from the 15th century) that ancient scholars developed a sense of respect and devotion for the classic

(a)   (b)

Figure 2.2 Hagia Sophia (6th century AD) in Istanbul. (a) External view and (b) buttressing system built by Sinan in the 16th century.

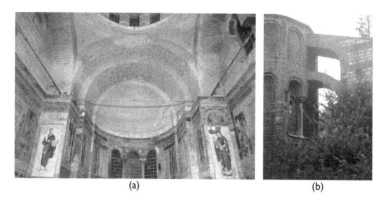

(a)   (b)

Figure 2.3 Strengthening of Chora Church (6th century AD) in Istanbul using (a) ties
and (b) flying arches.

arts (Greek and Roman) regarded as standard of excellence and sources
of inspiration but not as documents to be preserved. Moreover, although
without a mature approach to cultural heritage, Renaissance was also the
period of the first archaeological works and the onset of the first collections
of ancient art.

During the 15th and 16th centuries, although following the old style of
classic architecture, conservation of built heritage was carried in agree-
ment of the new architectural style. This was based on the classic ideal of
beauty defined in Vitruvius' *De Architetura* (around 29 and 23 BC), then
re-stated by Leon Battista Alberti (1485), as '*correspondence and proportion between all the parts of the building*'. As a consequence, interventions
on existing buildings must be perfectly integrated, that is, for the sake of
aesthetical harmonization, builders were employing elements related to the
primitive style of the construction (i.e. an *anachronism*), often resulting in
a mixture of architectural languages.

The façade of the Cathedral of Orvieto, Italy, represents a clear example
of this *retrospective approach*, where, still in 1786, Gothic and Medieval
elements were adopted to continue the construction started in 1310
(Figure 2.4a). The same trend can be found in the works of the 16th and
17th centuries to complete the Gothic San Petronio Basilica in Bologna
(even Michelangelo and Palladio were involved in the debate), even if the
church remained unfinished (Figure 2.4b).

Even though the retrospective approach represented a new attitude
towards conservation, Renaissance builders did not hesitate to demolish
abandoned ruins (valuable sources for elements and materials) or to substitute a construction for the sake of uniformity or fruition. For instance,
to reveal the increased power of the Catholic Church, Saint Peter's Basilica
(Rome, Italy, around 318–322 AD) underwent several radical interventions
until reaching the modern shape, completely different from the original one.

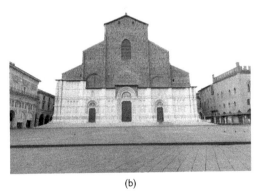

(a)   (b)

Figure 2.4 Retrospective approach in restoration. (a) Cathedral of Orvieto, started in
1310 and still under construction in 1786 and (b) San Petronio Basilica in
Bologna: despite the proposals in the primitive Gothic style, the construction
was left unfinished.

Given the extensive change, Saint Peter's Basilica can be considered as
an extreme case of conservation that opened a debate among scholars and
intellectuals of the time regarding the importance of the antiquity as a symbolic value. To many, in fact, the old, but original Saint Peter's Basilica was
still considered a tangible evidence of the perpetual history of the church,
thus more authoritative and influential than the new impressive building.
Accordingly, to preserve the evidence of the ancient basilica, in 1605, Pope
Paul V ordered to compile an iconographic inventory of the old ruins.

The classic ideal of beauty and the devotion for ancient Greek and
Roman, at the base of any Renaissance artistic expression, was called into
question during the subsequent centuries. Following the archaeological
excavation of Ercolano (1711), Hadrian's Villa (1734), and Pompeii (1748),
for the first time, the historical heritage was not seen as belonging to a
golden and undefined era, but with a clear awareness of the temporal coordinates. A clear distinction between the past and the present was marked,
with the past regarded as a concluded cycle encompassing cultures already
extinguished. In this regard, the historic-artistic value of the monuments
was directly linked to the representation of the ancient culture that produced it, thus worthy to be preserved and transmitted to future generations.
However, from the practical point of view, the interventions on historical
monuments were still performed according to the traditional way (*stylistic*,
that is, mimetic according to the primitive style), still far away from later
theoretical achievements.

Beside the artistic and emotional values of historical buildings, the first
attempts towards a 'scientific' approach were proposed by the intellectuals
of the Age of Enlightenment, with a new scientific interest for ancient and
oriental monuments. This led to a chronological and stylistic classification

of monuments (mostly Egyptian, Greek and Roman), underlining the need of respecting the different styles.

At the end of the 18th century, although the French Revolution (1789–1799) represented the beginning of a new political and social order, the consequent uprisings and insurrections led to a period of terror and devastation. During the conflicts, artworks were not spared, so that in 1794 Henri Grégoire (1750–1831), constitutional bishop of Blois and revolutionary leader, in order to put an end to this counter-revolutionary activity, coined the term *vandalism* and stated: '*Barbarians and slaves detest knowledge and destroy works of art; free men love and conserve them*'. Being also a member of the Commission of Public Instruction, in a series of three reports issued to the National Convention, he drew attention also to the conservation of manuscripts and the organization of libraries, as well as to the educational reasons for the conservation of cultural heritage.

Following Grégoire's reports, new decrees were drafted by the Commission of Public Instruction representing the first legal conservation measures. As an example, the recognition of the need for government intervention is to safeguard works of art even by coercion (2-year prison term for whoever damaged or destroyed artworks). Moreover, the government commissioned to expert institutions and individuals to reflect and decide about the way conservation should be carried out and what was worth to be preserved.

2.2 ONSET OF MODERN RESTORATION THEORY

During the 19th century, several theories and practices linked to different national movements in Europe started to develop: *archaeological restoration* in Italy (being Stern, Valadier and Camporesi the most influential persons), *stylistic restoration* in France (with Vitet, Mallerville and Viollet-le-Duc) and *anti-restoration* movement in England (with Ruskin and Morris). Although often extreme and contradictory, they represented the starting point for subsequent debates until achieving a synthesis in the most recent theories of restoration, which followed.

2.2.1 Archaeological restoration (Italy)

During the revolutionary years, the looting of artworks undertaken by Napoleon's troops exacerbated the emotional bond (born during Renaissance) to classical antiquities in Italy. In this regard, to preserve works of art, at the beginning of the 19th century, Pope Pius VII (1800–1823) committed the inspection and care of historical works of arts in Rome and Papal States to the leading artists of that time.

Regarding ancient monuments, the architects mostly involved in conservation were Giuseppe Camporesi (1736–1822), Raffaele Stern (1774–1820) and Giuseppe Valadier (1762–1839). In particular, given the very bad conditions of the Colosseum (the largest amphitheatre in the Roman Empire and eternal symbol of Rome), one of the first restoration projects involved this monument. After several inspections, Stern proposed to build a plain brick buttress to stop the lateral movement of the partly ruined eastern-side outer ring, forming a solid and economically feasible support in full respect of the architectural and historical values of the monument. Although the intervention (1806–1807) was criticized for the strong visual impact of the buttress, two aspects are worth noting: Stern adopted bricks as different and recognizable material and, by confining the deteriorated arches, he 'froze' deformation and cracks, creating a sort of snapshot of the instant before the pre-collapse condition (Figure 2.5).

In analogy with the approach to archaeological remains, Stern's idea was to preserve the building in the state as found (*'to repair and to conserve everything – even though it were the smallest fragment'*), giving birth to the so-called *archaeological restoration*. Accordingly, historical buildings should be archaeologically researched, scientifically studied and documented in detail. Only then it was possible to propose interventions avoiding any innovation (or 'creative' operations) regarding shape, proportions or decorations, except for removing later non-significant components. Moreover, monuments could be stabilized or completed by *anastylosis*, that is, the reassembly of ruined monuments from fallen or decayed fragments (including those found in excavations). New materials introduced to stabilize, or complete part of the monument, should be distinguishable and should have only essentialized (i.e. basic or absolutely necessary) decorations.

In the following years, the Colosseum underwent several works of maintenance, but a new intervention was requested for the stabilization of the western side (1824–1826). After Stern's death, Valadier proposed a solution somewhat different from his predecessor's: his project involved the

(a)   (b)

Figure 2.5 Stern's intervention on Colosseum. (a) Buttress for the eastern-side outer ring and (b) detail of a deteriorated arch, walled in and 'frozen' in the shape of an incipient collapse.

reconstruction of portions of missing structures *'imitating the antique even in minor details'*. Due to economic reasons (following also a French decree), he reconstructed arches, columns and cornices with bricks, then covered with a patina *a fresco* with the aim of resembling travertine. Given the higher cost, only bases and capitals of the columns were made in travertine. Although adopting the same shape of Stern's buttress, Valadier attempted to make his buttress as unobtrusive as possible and, due to the fresco imitation, a bit far from the concepts of archaeological restoration (Figure 2.6).

Another important monument restored by Stern initially (1818–1821) and, later, by Valadier (1822–1824) was the Arch of Titus, posthumously considered a perfect example of approaching historical built heritage. With the aim of restoring the original dimensions and proportions, they preferred to re-establish the arch rather than consolidating it. Using a strong centring, the original elements were carefully marked and dismantled one by one, then reassembled on a new brick core, and faced with travertine. Again, the choice of travertine was due to economical limitation, leaving the new part plain ('essentialized') in harmony with the original marble elements (Figure 2.7). Although criticized at the time for the supposedly larger costs and the loss of the historical value, the intervention represented an influential case for subsequent developments.

From a modern perspective, archaeological restoration is very close to today's current understanding. This approach, in fact, fostered respectful interventions based on archaeological research, avoiding over-interpretation and over-restoration. However, along the history, the theory often failed (as many subsequent theories) to recognize the chronological, stylistic and constructive complexity of many ancient constructions and the fact that they may have resulted from several historical additions or stages.

 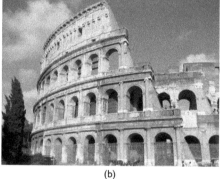

(a) (b)

Figure 2.6 Valadier's intervention on Colosseum. (a) Before (painting by Bernardo Bellotto, c. 1742) and (b) after the construction of the buttressing system.

(a)

(b)

Figure 2.7 Restoration of the Arch of Titus. (a) Before (painting by Giovanni Paolo Pannini, first half of the 18th century) and (b) after the intervention (with 'essentialized' reconstructed parts).

Along with Colosseum and Arch of Titus, after the fire of 1823, the Early Christian Basilica of Saint Paul Outside the Walls was badly damaged and in need of interventions. In spite of Valadier's proposal of completing the church in modern fashion starting from the surviving transept and apse, Pope Leo XII decided to rebuild the basilica in its earlier form with no innovations, unless for removing the effects of a ('capricious') intervention that modified the original construction. This represents one of the first examples of the so-called *stylistic restoration*, addressed next.

2.2.2 Stylistic restoration (France)

In Central Europe, the 19th century revealed a renewed passion for medieval art. Gothic architecture, in particular, became as an evocative and romantic source of inspiration (see, for instance, Goethe's description of Strasburg Cathedral in *Von deutscher Baukunst* – 'On German Architecture').

This is particularly evident in France, where Gothic architecture was regarded as a national style and also as an example of ideal rationalism. Accordingly, Ludovic Vitet (1802–1873), French General Inspector of Historical Monuments and pioneer of stylistic restoration, suggested that the interventions on monuments should be based on a deep knowledge of the original style to propose mimetic solutions grounded on evidences rather than hypotheses (adopting an *induction* process). Again, the definition of induction (logic) by *Oxford English Dictionary* can be synthesized as the process of inferring a general law or principle from the observation of particular instances (opposed to deduction). The lesson of Vitet was then extended by Prosper Mérimée (1803–1870), who added the concept of '*analogic criterion*': for the sake of coherence, it was possible to propose interventions on monuments according to the original style, taking inspiration from buildings of contemporaneous origin and the same region, but without rebuilding elements definitely lost.

Even if a new theory was about to be born, the restorers of the time were still doubtful and lacking any practical know-how, at least until the emergence of a new figure: Viollet-le-Duc (1814–1879), one of the most influential intellectuals of the 19th century. Starting from the lessons of Vitet and Mérimée, considering each building characterized by a certain, well-established style (involving specific forms and an architectural language), restoration was understood as *re-establishment*, that is, full recovery of the original integrity, as a sort of 'pristine' condition, even if it did not ever exist. This rather extreme concept allowed removing all the posthumous inclusions and restoring or (re-) building elements or portions of it through stylistic interventions based on analogical criteria following history of art and archaeology.

Although the strong conceptual statements, the first works carried out by Viollet-le-Duc were more moderate and characterized by interventions based on the full understanding of monuments (according to ancient treatises and documents of construction), still a key point of any restoration. In order to achieve this goal, in the early stage of his career, he even renounced to his personal ideas, stating the need of de-personalization of the restorer. This approach led Viollet-le-Duc to seek for the true original idea of the monument, paying the due respect to subsequent historical evidences and proposing solutions on a case-by-case basis (e.g. Madeleine Church in Vezelay, 1840–1859). In those years, architectural restoration became an autonomous discipline, where the structure and its rationale assumed a capital importance.

Over the years, the approach of Viollet-le-Duc became more and more radical, sometimes controversial and much criticized. In the restoration of the Cathedral de Notre Dame de Paris, 1845–1864, for instance, he undertook the construction of a spire that had never existed before to comply with his 'ideal' understanding of how a Gothic cathedral should be (Figure 2.8). Other exemplary cases include the restoration of Carcassonne walls (from 1852) and Pierrefonds Castle (from 1857). Although the architect focused on a careful survey of the ruins, he proposed a completely imaginary project. The intervention on Carcassonne walls is reported in Figure 2.9. Regarding the Pierrefonds Castle, for the sake of reviving the original symbolic and expressive value, he did not hesitate to propose an ideal model of medieval fortress, including every aspect of the building, from sculptures to furniture and wall decoration (Figure 2.10). In the case of Carcassonne, he even designed a brand-new church in Gothic stile inside the fortress (Saint Gimer Church).

Following Viollet-le-Duc's approach, all around Europe (although less emphasized), stylistic restoration gave way to interpretations, as well as personal and creative (even eclectic) solutions, achieving such an extent that

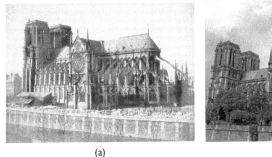
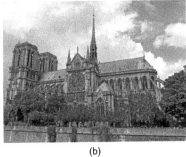

(a)   (b)

Figure 2.8 Viollet-le-Duc's restoration of Notre Dame de Paris in France. (a) Before (c. 1800) and (b) after the construction of the new spire over the crossing.

(a)   (b)

Figure 2.9 Viollet-le-Duc's restoration of Carcassonne walls in France. (a) Before (Auguste Mestral, 1851) and (b) after the intervention.

(a)   (b)

Figure 2.10 Viollet-le-Duc's restoration of Pierrefonds Castle in France. (a) Before (postcard, 1830) and (b) after the intervention.

later restorations have been needed to remove over-restorations and falsifications done by stylistic restorers. For the sake of achieving the primitive unity, restorers demolished many historically meaningful parts of emblematic monuments, deleting permanently evidences of the past.

The approach followed the assumption that the monument was mostly a work of art, neglecting its documenting value. This was modified by later restorers, who considered the monument also as a document, thus with a history worth to be preserved.

2.2.3 Anti-restoration movement (England)

If Gothic was considered the national and most rational style in France, the same style was perceived in England from the religious, political and moral points of view. This aspect, together with the devotion for nature, was the basis of the English approach to restoration in the 19th century when, after Augustus Welby Pugin (1812–1852), John Ruskin (1819–1900) became the most impacting person.

The interests of Ruskin, mostly a writer and art critic, were mainly focused on landscape painting and on the importance of nature observation as source of godly beauty. Accordingly, as far as architecture was concerned, he stated that it must be in harmony (symbiosis) with the natural environment, thus imitating the nature itself and its forms, with a clear predilection also for Gothic architecture and its natural essence. In his opinion, the external appearance was more important than structural or functional features.

In this regard, the approach of Ruskin to restoration was essentially based on the biological vision of the monument: it should be left to become a ruin or to die by itself when no further stabilization was possible. Accordingly, stabilization was accepted only to prevent failure, and conservation should consist only of maintenance through clear and recognizable interventions to avoid the need of further restorations. Works of the past should be kept intact (i.e. without subsequent action), because any intervention produces falsification and destroys the actual (natural) state, leading inevitably to a different building. In his words, the 'so-called restoration is the worst form of destruction: destruction accompanied with false descriptions of the thing destroyed'.

Although this approach is in extreme contrast with stylistic restoration, Ruskin stressed for the first time the importance of antiquity as a value, related to the age of the monument, inevitably marked by the ravages of time. Even though he did not have immediate influence among his contemporaries, Ruskin's proposal was essential for the development of the modern approach to the built cultural heritage, considered as a result of subsequent historical and natural events which made it unique. Ruskin's followers extended his concepts to the acceptance of the ruin as an intrinsic monument, where the operations for maintenance should be kept to the minimum possible. Moreover, the monument should be considered as an integral part of the landscape and urban context.

After Ruskin, one of the most influential intellectuals was William Morris (1834–1896), personally committed against stylistic restoration at the extent that he campaigned against and managed to stop the restoration proposed for San Marco's façade in Venice. His contribution to the restoration theory development can be summarized in the following points: (1) continuous maintenance is needed to avoid the need of restoration; (2) if restoration becomes needed, it may be acceptable, but an honest distinction must be made between ancient and new; (3) the importance and relevance of the historical and urban contexts of monuments needed to be reinstated; and (4) heritage is as wealth for the entire civilization. It is noted that Morris founded the Society for the Protection of Ancient Buildings in 1877, which is still active, to train new generations of building practitioners and craftspeople.

2.2.4 Philological restoration

After a period of debates involving, among others, the Italian Camillo Boito (1836–1914), in 1883, the 4th Conference of Architects and Civil Engineers of Rome gave birth to the first Italian Charter of conservation and the so-called *philological restoration*. This approach advocated a critical philological method based on the distinction between layers of intervention to present the historical structuring of buildings in their material authenticity. Note that philology is '*traditionally, the study of the history of language, including the historical study of literary texts*' (from Britannica), meaning the goal of reconstructing the original version of a building through critical analysis and interpretation. Apart from the importance of the statements, the Charter (edited by Boito) represented the first attempt to codify in a unique document the ideas and approaches of the intellectuals of that time. It was articulated in seven points (Jokilehto, 2018):

1. Architectural monuments should be consolidated (continuous maintenance) rather than repaired, and repaired rather than restored (in no case should a monument be left to die).
2. Should additions or renovations prove absolutely essential for the solidity of the structure or for other serious and unavoidable reasons, such additions or renovations should be executed in a different character from that of the monument.
3. Should the question be, instead, of constructing parts that have been destroyed or that for fortuitous reasons were originally never completed, it would be advisable anyhow that the additional or renewed blocks, whilst taking the original form, should still be made of obviously different material, so that no observer be misled. In monuments of antiquity and in others of particular archaeological interest, any part that must be completed for structural or conservation purposes should only be built with essentialized surfaces and using only the outlines of solid geometry.
4. In monuments, which derive their beauty, their uniqueness and the poetry of their appearance from a variety of marbles, mosaics and

painted decoration, or from the patina of their age, or from their pic-
turesque setting, or even from their ruinous condition, the works of
consolidation should be strictly limited to the essential.

5. Any additions or alterations that have been made to the original structure in different periods of time will be considered as monuments
and treated as such, except in the case that they are obviously infe-
rior artistically and historically to the building itself, and at the same
time, detract or obscure some important parts of it. Then, removal
or demolition of these alterations or additions appears advisable. In
all cases, feasible or worthwhile, the elements above should be preserved, either completely or in their essential parts, if possible, near
the monument from which they were removed.

6. Detailed reports should be produced along all stages of works, from
the initiation of even minor repair or restorations, then, gradually, of
all main stages of the work, and, finally, of the completed work to indicate clearly the parts that have been conserved, consolidated, rebuilt,
renewed, altered, removed or demolished. A clear and methodical
report on the reasons for the works and their progress should accom-
pany the documentation. A copy of the above-mentioned documents
should be deposited with the authorities responsible for the restoration of monuments or at the office in charge of restoration.

7. An inscription should be placed on the building to record the date of
restoration and the main works undertaken.

A few years later, Boito added the eighth point related to the dissemination
of the restoration works through publications.

Banning the idea of re-establishing the original integrity (as claimed
by the stylistic restoration), the conference pursued the preservation of
the monument as a historical document on past cultures, experiences
and arts (i.e. historical, spiritual and aesthetic values), without renounc-
ing to its contemporary (modern) use and needs. This can be done only
through an accurate and scientific approach combining respect towards the
remains and stabilization or re-composition based on a delicate and modern understanding of the building (i.e. Stern and Valadier's archaeological
restoration). Without any doubt, the difficulties posed by restoration as a
combination of so many issues, sometimes even controversial, requests to
the analyst carefulness, professionalism and judgement capacity.

2.2.5 Historical restoration

Given the objective difficulties of putting into categories such a complex and
delicate issue, at the end of the 19th century, besides the philological restoration, the *historical restoration* started to develop. Each building was considered as a unique and distinct case: a document whose different constructive
stages are to be recognized and preserved. With this view, each monument

needs specific criteria that are based on objective appraisal and documentary knowledge (old engravings, archaeology, history, detailed analysis of the existing construction, etc.), leading to the historical reality of the monument.

Although such a methodology was already contemplated in stylistic restoration, in the present case, the restorers did not accept the analogy criterion. From a modern point of view, the historical restoration can be seen as a combination of stylistic restoration with the rules of philological restoration, that is, authenticity for preserving the historical value. However, as already seen for other theories, the practical implementations of historical restoration were not successful due to inadequate interpretation of documentary sources, leading to excessive and erroneous restorations.

The Italians Luca Beltrami (1845–1933) and Gaetano Moretti (1860–1938), both Boito's pupils, can be considered the two main spokespersons of historical restoration. Conversely to their mentor, involved into the theoretical and conceptual development, they were mostly focused on practical projects. Beltrami, for instance, was the restorer of Sforza Castle in Milan between 1893 and 1905. Starting with a meticulous gathering of information about the original shape (including Leonardo's drawings located in the Louvre Museum in Paris), he proposed radical interventions resulting in an almost integral reconstruction of the castle, that is, stylistic restoration supported by an accurate documentation, which still entailed components that never existed (Figure 2.11).

On the other hand, Moretti was involved in the restoration of San Marco's Campanile (bell tower) in Venice, which collapsed on 14 July 1902. Although, in the first stage of the restoration, Beltrami proposed to rebuild it with the same materials and in the same place, Moretti underlined the need (supported by survey and calculation) of adopting modern techniques and materials, that is, steel and concrete. Regarding the Loggetta (Renaissance balcony, at the foot of the Campanile, by Jacopo Sansovino), also involved in the collapse, Moretti followed the same approach but with higher philological accuracy and technical competence.

(a)   (b)

Figure 2.11 Beltrami's (re-)construction of Filarete Tower of the Sforza Castle in Milan (it never existed before). (a) Before construction (1903) and (b) in 1956.

2.2.6 Scientific restoration

At the beginning of the 20th century, although the main theoretical concepts of restoration were already well established, the practical approach was slowly evolving, still adopting stylistic interventions.

In this scenario, following Boito's work, Gustavo Giovannoni (1873–1947) was the main representative of a new approach called *scientific restoration*. Accordingly, every monument was considered as a combination of history, art and technical skills, and an intervention would be a compromise between these elements. In this scenario governed by limitations, proposing rules, formulations and rigid schemes became impossible and only a deep understanding of the monument (on a case-by-case basis) could lead to a proper restoration.

With this aim, Giovannoni investigated the other theoretical trends, evaluating potential and deficiencies, and proposing restoration as a 'scientific' operation to preserve the monument and its (urban, natural, etc.) context as a contribution to identity and character. Additionally, he also focused on minor architectures, such as historical centres and urban volumes.

The position of Giovannoni can be considered as *intermediate* between archaeological (maintenance of the current conditions), stylistic, philological and, particularly, historical restoration. In this regard, he stressed the importance of a deep knowledge of history of art, with a particular consideration to the structure (structural members, construction techniques and ancient materials). In this complicated framework, he postulated a few requisites, as the one of minimal intervention and minimal additions, and, given the importance of the monument as document and artwork, no creative interventions should be acceptable. Only stabilization, *anastylosis* and technical re-composition (with technical additions) to ascertain preservation were allowed, even with new materials, especially reinforced concrete.

Moreover, according to Viollet-le-Duc, he considered the suitable utilization of the monuments as a fundamental part of conservation. In this regard, agreeing with the Belgian Louis Cloquet (1849–1920), he also adopted the distinction between *dead* and *living* monuments. Dead monuments (not necessarily ruins) belong somehow to the past, and they exist in the present as evidences of ancient cultures and an extinguished time, that is, as pure art documents. For this type of monuments, for example ruins, castles, fortresses, he envisaged stabilizing interventions. On the other hand, living monuments do not belong to the history but they are still in use, and this cannot be isolated from the artistic and aesthetic values. Temples, churches and assembly buildings should be restored and updated for new use (possibly close to the original one), still preserving the importance of all the parts (that is, an historical/philological approach).

Well aware of the difficulties posed by each case, Giovannoni proposed five typologies of interventions:

1. consolidation: re-establishment of bearing capacity through technical resources;
2. re-composition: assembly of the dislocated parts by *anastylosis*, provided that the new additions are noticeable;
3. liberation: removing all the unstructured parts;
4. completion: limited additions;
5. renovation: transformation of existing elements, and construction of new ones considered as essential.

Despite the theoretical contribution, Giovannoni's works were based more on innovation than conservation. Still, in the first decades of the 20th century, he was a prominent figure in the restoration field, being one of the main promoters of the Athens Conference (1931) and Athens Charter (1931), as well as of the Italian Restoration Charter (1932).

2.3 THE ATHENS CHARTER AND USE OF MODERN MATERIALS

The need of an efficient and international support for the preservation of built cultural heritage set the framework for the First International Congress of Architects and Technicians of Historic Monuments held in Athens in 1931, involving almost 100 experts from around 20 different countries. After debates and reports, often in line with the approaches of the previous decades (Boito and Giovannoni's particularly), at the end of the conference, the Athens Charter was approved and, like the Italian Charter of 1883, it was articulated into seven points. As major contributions, it opened up international cooperation, and it was legally adopted (although the practice still contained more or less pronounced stylistic traces).

The main tendency during the conference was to prevent future damage by initiating a system of regular maintenance that would limit the future need of repairing or strengthening actions. In case restoration appears to be indispensable, the historic and artistic work of the past should be respected, without prescribing the style of any given period. Moreover, the Charter recommended the respect of the ancient context close by the monuments. In case of ruined structures, the method of *anastylosis* should be implemented, and any new material should be recognizable. Finally, international cooperation was emphasized to strengthen the protection of historic works of art.

The conference also considered the use of new materials, iron and reinforced concrete in particular, more and more frequently adopted to simplify expensive interventions, rather than considered necessary for the given condition of the monuments. After the Athens Charter, new materials, together with the means and procedures of modern technology for both consolidation and re-integration (see also Italian Restoration Charter promoted by Giovannoni in 1932), became the default practice, without argumentation or discussion (often leading to detrimental interventions).

With respect to the use of iron (and steel), recognized restorers of the past had already addressed the subject. Valadier, for instance, rejected the use of iron because it would reduce the strength of structures when corroding; in 1809, the Italian Luigi Bardet (military engineer) harshly criticized the use of iron because, when corroding, it increases in size, damaging more than preserving. In France, Mallerville opposed to the use of iron in the consolidation of Rouen Cathedral, and it is only with Viollet-le-Duc that iron was partially accepted, notwithstanding the concerns related to its durability and mechanical compatibility with traditional materials.

In spite of this awareness, over time, iron and steel became widely used in restoration causing an enormous amount of damage and conservation problems due to corrosion, requiring, in turn, subsequent repair or 'restoration of restorations'. A typical example is the metallic strengthening of the Parthenon in Athens (1870–1872), recently removed (Figure 2.12a). Nowadays, rather than iron, titanium can be used for cramps/dowels (Figure 2.12b), together with other durable and lower cost materials such as stainless steel or fibre-reinforced polymer rods, due to their better resistance to corrosion.

On the other hand, as stated by Sir Arthur Evans in 1925 regarding the works on Knossos Palace, the use of concrete for all types of construction members opened a new age for restoration and conservation. This was the view of Nikòlaos Balànos, for whom reinforced concrete was respectful, solid and durable. In many monuments, from the French Gothic cathedrals to Aztec

(a)   (b)

Figure 2.12 Restoration works in the Parthenon. (a) Stone members taken down to remove iron strengthening and (b) modern use of titanium.

remains in Mexico, from Roman to Egyptian temples, Portland cement and concrete were used, generating significant deterioration due their incompatibility (mechanical, chemical, rheological and thermal) with original materials, requiring, in turn, much repair or re-restoration works. In addition, it must be stressed that reinforced concrete (and even plain concrete in some cases) has limited durability for restoration purposes, normally resulting in an irreversible intervention. Details of damage mechanisms are to be found in Chapter 6.

As an example, Figure 2.13 shows the use of abundant metallic rebars in the reconstruction works undertaken on Temple C at Selinus in Sicily (Italy), and the effect of corrosion on a lintel of the Acropolis of Rhodes in Greece. Besides the irreversibility of the intervention, one of the main drawbacks of concrete lies on carbonation and the subsequent corrosion of the bars. The expansion of corroding metals creates severe tensile stresses in the concrete, which cause cracking, delamination and spalling, that is, the expulsion of concrete cover. This, in turn, makes the bars more exposed to corrosion, leading to a reduction of the cross-sectional area and, in the absence of repair, to total loss.

Excessive use of cement in monuments and poor workmanship on rare and important archaeological structures is not unusual. Figure 2.14 shows two examples of how modern reinforced concrete, cement mortar patches, bricks, cement mortars and plasters (all newly introduced materials to the archaeological area) could exceed the and materials. Together with the corrosion of the reinforcing bars, the extensive use of unauthentic materials introduced by foreign experts compromised the integrity of the monuments, and created a serious precedent for local professionals in charge of undertaking repairs and maintenance of the site.

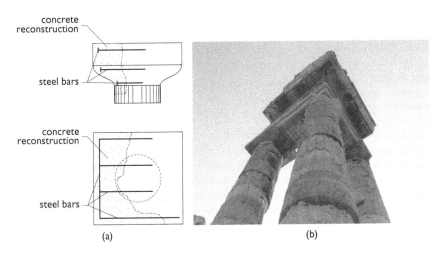

Figure 2.13 Use of metallic bars for lintels. (a) Reconstruction of a capital at the Temple C at Selinus in Sicily (plan and elevation) and (b) corrosion and subsequent concrete spalling at the Acropolis of Rhodes in Greece.

(a)   (b)

Figure 2.14 Two examples of inadequate use of cement in restoration works. (a) Temple of Venus and Roma at Rome (Italy) and (b) Villa of the Emperor Hadrian at Tivoli (Italy). Note how cement patches exceed the original materials used by Romans.

2.4 FROM THE SECOND WORLD WAR TO PRESENT

The massive destruction of monuments caused by the Second World War produced a sense of deep discouragement and confusion among the intellectuals involved into the conservation of cultural heritage. In 1944, when the conflict was still ongoing, the Italian Roberto Pane (1897–1987) described well this feeling: the preservation of monuments ceased to be a practice related to cultural and artistic values, becoming a need for saving the remains, even at the cost of disregarding the concepts (and the achievements) of the established restoration practice. Despite the principles of the Athens Charter, monuments were stabilized, repaired or even reconstructed (practice widely employed) with no scientific or historical insight. Moreover, due to economic reasons, reinforced concrete was widely adopted. In other words, the war brought the scientific restoration into deep crisis, considered inadequate and inapplicable to face such amount of destruction (as stated by Renato Bonelli in 1966).

The reconstruction was performed according to the philological approach, but most of the times freely interpreted and disregarding the principles of minimum intervention and distinguishability of modern works. In this complicated scenario, Guglielmo de Angelis d'Ossat (1907–1992) proposed in 1948 the following principles, according to the level of damage:

1. light (damages at the roof, cracks of modest width):
   • 're-establishment' (integral recovery);
2. severe (roof destroyed, large cracks in the walls, partial failures, disconnection between the parts of the building):
   • reconstruction (philological intervention, even though not minimal);

- update with respect to the current condition, due to significant losses or because the damages revealed a previous configuration of the monument that is worth to be maintained. The Victory Gate (*Siegestor*) in Munich belongs to this category, where, as a symbolic scar of the war, a simplified restoration of the inner side of the gate was proposed with the inscription 'Dedicated to victory, destroyed by war, urging peace' (Figure 2.15);

3. total (almost destroyed):
   - re-composition by *anastylosis* or renounce to restoration: Kaiser Wilhelm Memorial Church (Berlin), for instance, was not rebuilt as a reminder of the war (Figure 2.16), whereas the ruins of the old Saint Michael's Cathedral in Coventry (West Midlands) were integrated in the project of the new cathedral, without reconstruction (Figure 2.17);

(a)   (b)   (c)

Figure 2.15 Victory Gate (Siegestor) in Munich. (a) 1891, (b) after the Second World War and (c) current view.

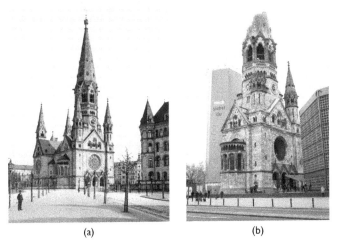

(a)   (b)

Figure 2.16 Kaiser Wilhelm Memorial Church in Berlin. (a) Before the Second World War and (b) today (not rebuilt as a reminder of the war).

(a)   (b)

Figure 2.17 Saint Michael's Cathedral in Coventry (West Midlands). (a) Ruins after the Second World War and (b) current aspect with the ruins integrated into the new cathedral as an open space.

(a)   (b)

Figure 2.18 Historical centre of Warsaw in Poland. (a) 85% destruction during the Second World War and (b) subsequent full reconstruction.

- full reconstruction in the case of very important monuments, due to psychological or symbolic values (as already done for the San Marco's Campanile rebuilt after the collapse of 1902). The historical centre of Warsaw represents the most striking example of the importance of restoring the image of the city as it was before bombing (Figure 2.18).

As already stressed, the urgent and extraordinary measures required, although commendable for the efforts of addressing the (material and psychological) damages of the war, marked inevitably the end of the scientific restoration, that is, Boito and Giovannoni's approaches, as well as the Athens Charter. In this intense cultural environment, many intellectuals started to criticize the principles of the Charter, paving the way to a new approach. Beside the 'minimum intervention', criterion usually disregarded, the scientific restoration revealed itself inadequate in dealing with monuments mainly through historical arguments, while not providing sufficient importance to artistic criteria. Moreover, the Athens Charter considered valuable only the ancient context close by the monument, neglecting urban environments and landscapes as testimonies of past cultures.

In 1950, Pane stated that each monument should be seen as unique, with its distinct values and its genuine and specific restoration problem. Pane also stated that restoration of an artwork is an artwork itself, whose aesthetic qualities cannot be lost due to distinguishable additions (as suggested by Athens Charter). Accordingly, restoration projects should care for aesthetics and functionality, while not forgetting about historical issues. In this fervent debate, Bonelli added the concept of spiritual and expressive value in 1959, according to which the restorer must operate to identify, reintegrate and preserve the artistic qualities of the monuments by means of a critical and creative process.

All these stances, together with a harsh refusal of the approach adopted during the post-war reconstruction, led to the so-called *critical restoration*, mainly proposed by Pane, Bonelli and Cesare Brandi. Given the extreme variety of cases and the need for safeguarding a large amount of monuments, restoration could not be constrained within stiff (scientific) boundaries. Therefore, since each monument has its distinct values and poses specific restoration problems, no general concept on restoration is possible. On the other hand, only a historical and critical understanding of the monument (which overpasses the philological knowledge), together with the sensitivity for aesthetics and historical intelligibility, can guide the restorers, allowing even re-integration of missed parts and new additions.

The need of an international debate led to the First Congress of Architects and Specialists of Historic Buildings held in 1957 in Paris, where all member states of the United Nations Educational, Scientific and Cultural Organization (UNESCO) joined the International Centre for the Study of the Preservation and Restoration of Cultural Property (ICCROM). The Second Congress, held in Venice in 1964, adopted 13 resolutions, of which the first one represents the International Restoration Charter, better known as Venice Charter, and the second one, put forward by UNESCO, provided for the creation of ICOMOS, which currently brings over 10,000 individual members from 153 countries, 320 institutional members, 110 national committees and 28 international scientific committees (last update May 2017). This is a global non-governmental organization dedicated to the application of theory, methodology and scientific techniques in the conservation of the built cultural heritage. The objectives of ICOMOS are to

- bring together conservation specialists from all over the world and serve as a forum for professional dialogue and exchange;
- collect, evaluate and disseminate information on conservation principles, techniques and policies;
- extend the influence of the Venice Charter by creating flexible doctrinal texts for specific sectors of built cultural heritage;

- organise and manage expert missions at the request of heritage administration and legal entities;
- put expertise of highly qualified professionals and specialists at the service of the international community.

ICOMOS also plays a vital role in counselling UNESCO on the cultural properties to be possibly included in the World Heritage List and on the reporting of the state of conservation of the properties already listed.

At the Congress on the European Architectural Heritage held in Amsterdam in 1975, the European Charter of the Architectural Heritage was proclaimed and then adopted by the Council of Europe. After stressing the importance of regular buildings in the old towns (often referred to as 'minor heritage') and the importance of villages in their natural or man-made settings as built cultural heritage, the Charter focused on the current conditions of monuments. Article 6 reads:

> (Architectural) heritage is in danger. It is threatened by ignorance, obsolescence, deterioration of every kind and neglect. Urban planning can be destructive when authorities yield too readily to economic pressures and to the demands of motor traffic. Misapplied contemporary technology and ill-considered restoration may be disastrous to old structures. Above all, land and property speculation feeds upon all errors and omissions and brings to nought the most carefully laid plans.

In the following years, the Italian Restoration Charter of 1972 and the subsequent update in 1987 are worth noting. In particular, the former introduced the concept of reversibility of restoration (Art. 6) as a way of not limiting future interventions, and the authorization by the Central Restoration Institute for the adoption of new procedures and materials (Art. 11). The latter, following the National Conference about Restoration, held in Rome in 1986 (organized by Paolo Marconi and Corrado Maltese), for the first time clearly stated that reinforced concrete should not be used systematically to solve restoration problems. Traditional and historical techniques and materials should be preferred. An extract of Article 7d reads:

> The experience of the last years has shown us to mistrust hidden additions (insertions) of materials such as steel, reinforced injections with Portland cement or resins, due to their invasiveness, limited durability, irreversibility and limited reliability. Hence, traditional consolidating measures are preferable as they are easily controllable and replaceable.

Considering the stability of monuments, Bruno Zevi (1918–2000) is probably one of the first restorers (long after Viollet-le-Duc) who stated, in 1957, the need of respect towards the structure and its resisting principles: each monument has a static history which should not be altered. Similarly, in 1988, Stefano Gizzi stated that each monument has an internal history (besides the external one), which coincides with the history of its structural behaviour and its static model, which should not be altered. However, although the theoretical concepts were defined, they were not adopted during most part of the 20th century.

The first document that specifically dealt with structural aspects of conservation was issued at the symposium on Structural Aspects of Restoration of Architectural Heritage organized in Ravello (Italy) in 1995, with delegates from ICCROM, ICOMOS and International Association for Bridge and Structural Engineering. One year later, Giorgio Croci (one of the editors of the symposium) promoted the formation of International Scientific Committee for the Analysis and Restoration of Structures of Architectural Heritage (ISCARSAH) becoming its president until 2005. The ISCARSAH Recommendations for the Analysis, Conservation and Structural Restoration of Architectural Heritage (following the ones of the Ravello Charter), detailed in Chapter 1, were agreed at the Paris meeting in 2001. The Principles (first part of the Recommendations) were adopted at the ICOMOS General Assembly in Zimbabwe in 2003.

2.5 OPEN ISSUES

The history of construction, which coincides with the history of civilization, dates to about 10,000 years as stated earlier. In that perspective, conservation and restoration is a rather young science, with important developments in the last 100 years. Debates have generated very different theories (e.g. from stylistic to critical restoration), all of which have contributed, to a lesser or larger extent, to the modern understanding. In general terms, stylistic interventions are nowadays universally rejected and even legally prevented, limiting the amount of 'innovation' and 'creation' to the minimum. In turn, 'passive fatalism' is still influencing and has a number of supporters.

In this scenario, two important issues still pose major challenges. The first issue regards reconstruction in case of sudden destruction caused by natural disasters or human actions (e.g. wars). It is possible to accept a scientific and careful reconstruction of the monument, as for the reconstruction of Noto Cathedral (Italy) and Mostar bridge (Bosnia-Herzegovina), and the proposed projects of Bam Citadel (Iran) and Bamian Buddhas (Afghanistan), but this is not a universal solution (photos of the mentioned monuments are shown in Chapter 6). An important recent reference is the 'ICOMOS Guidance on Post-Trauma Recovery and Reconstruction for World Heritage Cultural Properties' document (ICOMOS, 2017).

The second issue concerns the conservation of the built cultural heritage of the 20th century, and the Modern Movement in particular, built with less durable materials, such as metals and reinforced concrete. The conservation principles normally adopted for ancient buildings are hardly applicable, and additional efforts and research are needed (ICOMOS, 2014).

Chapter 3

Construction materials and main structural elements

In the prehistoric era, with the need of freely roaming for the purpose of hunting, the ancestors of man used to dwell in occasional caves or temporary tents. The latter were made by leafy branches leaning against the trunk of a tree, or by branches leaning against each other, creating the inverted V-shape of a natural tent (circular or oval in-plan), with the bottom of each branch supported by bones, stone or local materials to hold it firm on the ground.

It is with the onset of agriculture (some 10,000 years ago) that nomadic life ended, and man started to build permanent shelters and houses. The tent-like structures evolved into round houses, giving birth to early villages. In this regard, the city and the fortified walls of Jericho (8000 BC) are usually addressed as the first example of masonry construction. Since then, according to the increasing needs of the society, construction materials, technology and techniques have restlessly developed, marking every period with extraordinary examples of human ability.

With no intention of presenting an exhaustive revision on this topic, for which an unlimited literature exists (e.g. Davey, 1961; Allen and Iano, 2013), this chapter introduces the reader to the main structural materials adopted along the history, with particular reference to masonry and timber, the associated technology and their main uses. The chapter also addresses, even if briefly, the development of construction materials up to the end of the 20th century, that is, when steel and concrete became the main protagonists of structural engineering. In this regard, metals are discussed mostly in combination with timber and masonry, with few references to bridges and buildings of the 19th and 20th centuries.

3.1 MASONRY

With the increasing necessity of taller buildings and more resistant dwellings, man started to process the raw materials that could be found nearby, most probably stone, mud, reed and timber. However, according to the location, only some of these materials were easily available. Only in a later stage, the evolution of civilization and the development of transportation means led to the use of materials according to their specific characteristics, the type of building (public, military or housing) and the wealth of the owner. In this regard, stone was considered the most durable material, whereas earth was often used where stone was scarce.

Placing stones, clay bricks or adobe units on top of each other, laid dry or bonded with lime mortar, creates masonry, which is possibly the first, but also one of the most used, durable and efficient construction techniques of all times. Masonry can be built with a single material or with a mixture of materials, with several layers of different constitution and elements of different sizes. Despite the simplicity of this manufacture process, masonry can also be considered as the most resistant construction technique of the past. In this regard, masonry elements were mainly aimed at enclosing space (providing shelter from sight, wind, rain and temperature) and at bearing loads (e.g. from floors and roof systems, but also the masonry elements themselves).

The strength of masonry depends mostly on the strength of the units, even if mortar also plays a role. Historical masonries may have the following characteristics: adobe, clay bricks or stone blocks with mortar to bind units all together, or dry-joint (no mortar) stone masonries, which rely only on the friction between stones to stay stable.

In the following, stone, earth, brick and mortar, and the main structural elements made with these materials are discussed. In particular, except for lintels and vaulted structures (discussed in Chapter 4), walls, columns and foundations are addressed next.

3.1.1 Stone

Stone represents the strongest, most used and most durable material of the past, usually preferred for structures of the greatest importance. As a matter of fact, given the larger strength, stone masonry is also the one that better survived along the history, representing nowadays an incredible evidence of prehistoric man. To many, the oldest known man-made structure is represented by the wall that blocked two-thirds of the entrance to the Theopetra Cave in Greece, datable around 23,000 years ago (probably built to protect its residents from cold winds at the height of the last ice age). In turn, Göbekli Tepe (southeastern Anatolia Region of modern-day Turkey) is the oldest known example of monumental architecture (9600 BC), built probably for ritual or religious purposes. It consists in dozens of massive cleanly carved limestone pillars arranged into a set of rings, with bas-reliefs of animals.

Besides the clear advantages of stone, the ancient builders needed to solve two main difficulties, namely, extraction from the quarry and dimensioning (in case it was not possible to use blocks as found). Both are related to the hardness and internal structure of stone, in turn intimately related to the geological process that the rock has suffered. Without entering into the merits of rock classification, for which the reader is referred to devoted literature (e.g. Siegesmund and Snethlage, 2014; Schön, 2015), rocks are generally divided into igneous, sedimentary and metamorphic.

Igneous rocks form through cooling and solidification of magma or lava. With the exception of tuff (produced by consolidation of volcanic ash and stone, largely adopted by Romans), these rocks are hard, and thus difficult to be extracted and worked. This also led to their restricted use, limited to public buildings (due to the larger durability), and to the maximum simplicity of the forms. Only with growing experience and better tools it was possible to handle this type of material. Regarding igneous rocks, granite represents the main one and was extensively used in Mesopotamia and in the Valley Temple of Chephren at Giza, in Egypt.

Sedimentary rocks are the result of accumulation and cementation of fragments of earlier rocks, either on land or under water. Sandstone and limestone belong to this group (the former being usually harder than the latter), and they can be sawn more easily than igneous rocks. Moreover, since the accumulation followed a layer-by-layer deposition, they can easily be split in the direction of the bed, thus facilitating the carving in square shape. According to the quarry, blocks may be very large for lintels, columns or monolithic shafts, or smaller for other different construction purposes. Luxor Temple and the Abu Simbel colossal figures of Ramses II at Great Temple are two magnificent examples of the use of sandstone. Romans used travertine extensively, which is a form of limestone.

Finally, metamorphic rocks are formed by subjecting any rock type to different temperature and pressure conditions than those in which the original rock was formed, resulting in a profound change in its physical and chemical properties. In addition to schist, which is a foliated rock (i.e. composed of or separable into layers), marble is the one of the largest structural importance and it derives from limestone. Greeks and Romans made extensive use of marble for building entire temples or simply to clad masonry constructions.

From the mechanical point of view, no matter the geological process, stone exhibits a good compressive strength, but a low tensile and shear strength. These peculiarities restricted the use of stone to vertical load bearing elements (walls, piers and columns) or to build lintels for doorways or passageways by means of large blocks (to ensure that tensile stresses are limited and arching effect can develop within the element, see Section 4.2.1). Only in a later stage, with the invention of the arch, stone was adopted to cross larger spans (e.g. in bridges, vaults and domes).

Another essential consequence of the mechanical properties of stone regards the importance of guaranteeing a uniform stress distribution (only compression) between the units. This can be obtained either by careful dressing of the bed joints (according to the workability of the stone) or by interposing a thin layer of another material. In ancient times, the builders

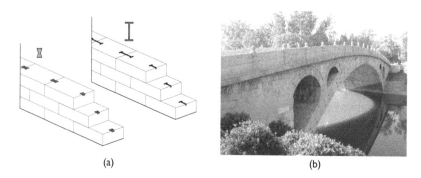

(a)   (b)

Figure 3.1 Use of iron connections in masonry structures. (a) Schematization of the construction of Greek temple walls and (b) Anji Bridge, China (605 AD).

preferred to dress the blocks, and the thin layer of mortar was probably used more as a lubricant for manoeuvring the blocks than for bearing purposes. Also, sheets of lead were used to level the courses, as for the piers of Hagia Sophia in Istanbul (6th century AD).

In order to limit shear stresses on the interface between the units, the joints should be ideally at right angles with the force resultant (compression). However, in case the direction of the action was expected to vary (e.g. earthquakes or soil settlements), together with the limited resistance provided by friction, several expedients were adopted, namely, dowels, cramps and dovetails. For instance, the seismic performance of the walls of Greek temples was improved by connecting the blocks by means of iron cramps embedded (to avoid corrosion) in lead (Figure 3.1a). Iron cramps were used also for the construction of Anji Bridge (also known as the Zhaozhou Bridge), the oldest purely stone segmental arch bridge (and the oldest standing bridge in China) built in 605 AD by Li Chun (Figure 3.1b). The bridge has a span of 37 m, made of limestone blocks. In general, iron connections were used to locate lintels or drums of multidrum columns (drums are cylindrical blocks that form the shaft of a column), as well as to bond blocks across joints in walls and foundations.

3.1.2 Earth and brick

Although the durability of earlier earthen rudimentary buildings was low, the simplicity and quick availability of the material made earthen construction one of the most used ancient construction technologies, which was subsequently adopted mainly where stone was not available. The first shelters using earth were probably made of brush and small wood members covered with mud for waterproofing. However, it is worth noting that earth can only be used for construction purposes if it has inherently good cohesion, provided by the presence of clay, which acts as a natural binder.

Clay belongs to the geologic group of sedimentary rocks, which are created by the accumulation of sediments along thousands of years. This process provides clay with the properties which also make it appropriate for brick manufacture (red clay, most common) or for pottery and ceramic wares (typically made by kaolin). Clay can be found in the so-called *clay pits*, mostly located close to riverbeds, lakes and valleys, as for the case of alluvial valley of Nile or in Mesopotamia area. In these regions, in fact, earth (in the form of mud with the addition of gravel, sand, straw or reed) has been continuously used for millennia in the construction of simple dwellings.

Being basically composed of soil and water, the drying process is necessary for the earth material to get strong. However, given the consequent shrinkage, the elements undergo extensive cracks, only partially limited by the straw added to the mixture (similar to modern fibre-reinforced concrete). To prevent this phenomenon, ancient builders started to mould small elements, so-called *adobe*, which were sun-dried before being used.

The building methods varied according to locations and cultures, and the most ancient ones can be synthesized as follows:

- direct digging: the first habitations were built directly in the ground by digging out a layer;
- straw-clay: clayey soil was added to straw (the clay binds the straw together), and it was used to build several building components (bricks, blocks, panels), in a sort of first ancestor of *ferrocement* of the 19th and 20th centuries;
- wattle and daub: a bearing wooden structure was filled with clayey earth mixed with straw, to prevent shrinkage;
- direct shaping: an ancient technique that made use of a very plastic earth to model forms directly without using any kind of mould or formwork.

Among the different techniques, two seem to have had a larger dissemination and are extensively used until today, namely, adobe and rammed earth. The former consists in sun-baked earth bricks, made by a mixture of earth, sand and straw shaped by hand in wooden or metal moulds; the latter is earth compacted in formwork with wooden forms and rammers. Adobe of high quality and regular size started to be used from at least the third millennium BC, allowing the construction of massive walls, ziggurats, pyramids as well as vaults and arches. In spite of its very old origin and the overall poor materials adopted, these techniques are still used today. Today, earth construction has reached high scientific and technological levels, thanks to its good acoustic and thermal properties, low cost and low embodied energy and carbon. However, earthen constructions have a low resistance to earthquakes and must be protected from rain (and moisture in general).

Despite the structural improvements, simply sun-dried earth bricks were still vulnerable to climate effect, especially alternate wetting and drying. If bricks are not fired, in fact, the process of dehydration is reversible and they can reabsorb water. Only a temperature of firing around 1,000°C produces irreversible transformations through permanent chemical changes. Probably, following the observation that bricks near a cooking fire, or bricks remaining after a thatch roof burnt, were stronger and more durable, the practice of burning bricks became common. Given the highly expensive process, fired bricks were limited to the elements for which sun-dried bricks were inadequate, such as the outer leaf of walls, or roofing tiles, becoming a major structural material only during the Roman Empire.

Traditionally, the manufacture of clay bricks could be divided into four stages. After the extraction of raw clay from pits, the clay was accumulated and moved to an open-air storage area. During this period, the raw material was rummaged to reduce impurity and soluble salts to a minimum, leading to an overall more homogeneous material. Despite the undisputable preference for materials available nearby, the selection of raw material was essential to improve the performance and durability of bricks. Subsequently, clay was further crushed and mixed with water (*tempering*) to obtain a mix of adequate plasticity to facilitate moulding (in square or rectangular shapes). The moulds were generally in wood and bottomless, generally prepared directly over the soil, with, in some cases, a thin layer of sand to ensure a low bond to the substrate.

The mixture was then left to dry in shelters for several days according to climate conditions and, if possible, to avoid early warping and cracking, the bricks were protected from direct sunlight. Finally, to gain more resistance, the bricks were fired in kilns or left to dry at the sun (in hot climate regions). In the former case, an adequate supply of fuel was necessary, mainly wood (coal was extensively adopted only starting by the end of the 19th century). The fact that this supply was not always available may have partially accounted for the continued use of sun-dried bricks in the Near East.

Compared with stone, bricks present three main differences in history: (1) the compressive strength is strongly dependent on manufacturing and is generally lower than stone; (2) although produced in several shapes and sizes, bricks were usually smaller than stones; and (3) the shape of bricks, even though fairly uniform, presented irregularities in size that often requested a relatively thick mortar joint, which provides a lower overall compressive strength (see Section 3.1.3). In turn, the small size of bricks made the construction of curved elements (arches and vaults) easier by slightly tapering the joints.

The evolution of brick dimensions followed the evolution of civilizations very closely. In the Roman period, bricks were mostly large squared tiles, with a thickness of 5–6 cm and up to 40–60 cm of edge (dimensions based on Roman feet size). The dimensions of clay bricks evolved, naturally, to values more easily handled, such as $(25 \div 35) \times (10 \div 20) \times (4 \div 6)$ cm^3.

In Tuscany, for instance, during the Middle Ages, the bricks were $(29 \div 30) \times (12 \div 13) \times (4.5 \div 6)$ cm³ large. Similarly, in Bologna, the bricks were $28 \times 13 \times 6$ cm³, reducing the thickness to 5 cm only starting from the 16th century. Note that a width slightly larger than 10 cm is optimal to allow handling with only one hand.

3.1.3 Mortar

In the broadest sense, mortar is a workable paste used to bind stone or brick units together and build masonry elements. In particular, eliminating the effect of irregularities in the units, mortar facilitates their stacking and prevents the concentration of stresses, thanks to a more uniform distribution of compression stresses. The early mortars included asphalt or clay; only in later stages, mortar assumed the modern meaning of a plastic mixture of sand, water and binder (lime or cement) in proportions able to guarantee a good balance between workability (in fresh condition) and strength (in hardened condition).

Probably the first natural deposit of cement compounds dates back to 12 million years ago, when the earth itself was undergoing intense geologic change, with the reactions between limestone and oil shale during spontaneous combustion in Israel. In China, there are evidences of cementitious materials used to hold bamboo together for building boats as well as in the construction of Great Wall (3000 BC). In the same period, Egyptians used gypsum and lime mortar for building the Pyramids. From the chemical point of view, these earliest mortars were non-hydraulic, that is, they need exposure to the carbon dioxide available in the air (hence, the name 'aerial') to harden. Thus, such mortars cannot be used under water (or in high damp environments), and they take a considerable time to harden inside a thick wall, usually requiring slow construction.

Gypsum is a very common mineral associated with sedimentary rocks, whose deposits are found in lakes, sea water and hot springs from volcanic vapours. In order to be usable, the gypsum stone was reduced into powder by burning it at a temperature of 160°C. The lower firing temperature and the quicker setting up with respect to the lime mortar may be the reasons why gypsum mortar was the typical mortar in ancient brick arch and vault constructions. Nonetheless, gypsum mortar is not as durable as lime mortar in damp conditions. Lime mortar is obtained by crushing and burning limestone at high temperatures in lime kilns. Given the requested large amount of fuel, lime was primarily used in Europe, where great supplies of wood and limestone were available.

From the mechanical point of view, the compressive strength of mortar is low (usually much weaker than clay bricks or stone blocks), even if, under triaxial compression, high strengths can be obtained. This characteristic considerably affects the overall masonry strength, which gets lower as the mortar joints become thicker (or the confinement of mortar joints reduces).

Mortars have not faced significant changes along the history, at least up to the 19th century when Portland cement was introduced. With no doubt, mortar and concrete suffered an important breakthrough during the Roman Empire when it was widely adopted, also for building masterpieces of architecture such as the Pantheon and Colosseum. The concrete was called *pozzolana*, whose designation is derived from one of the primary deposits of volcanic ash, that is, Pozzuoli, Italy. It allows to develop hydraulic mortars, which react directly with lime independently of air contact. Adding natural or volcanic ash (i.e. a compound of silica or alumina) resulted in higher strength, faster strengthening and, overall, more durable mortars. The technique of adding fly ash (natural or from burning wood) or other pozzolanic materials (e.g. crushed underfired bricks, tiles, or potsherds) became eventually widely disseminated.

Not much is known about the proportion of mixtures: Pliny the Elder reported a mortar mixture of one part lime to four parts sand (in volume). Vitruvius, instead, reported two parts pozzolana to one part lime for the binder. Other bizarre ingredients were used as admixtures, such as animal fat, milk, blood and even wine to slake the lime (i.e. to obtain a paste of calcium hydroxide to be used for mortars), and also soap (for waterproofing effect of mortars) or various natural polymers (e.g. for increasing the bond), with many different purposes.

3.1.4 Walls

As far as walls are concerned, each geographical region and period in history showed different ways of building. As discussed earlier, several materials were available, but it is only with the use of stone, adobe and clay bricks that a relevant development was achievable. The main goals of ancient builders were usually to dress the units in such a way they could fit closely, with less amount of mortar, avoiding aligned vertical joints.

The designation adopted for the earlier building technique is 'megalithic' and derives from the dimensions of the units used. Two examples of megalithic masonry systems are shown in Figure 3.2. In the first case, the rubble irregular stones (with minimal dressing) led to joints rather wide and units often working in bending due to uneven bearing. In the second case, instead, the dressing is still minimal, but closer joints guarantee a better performance of the wall.

Megalithic polygonal stone masonry was widely used by Egyptians and by Incas in Peru (Figure 3.3). For instance, the Saqqara pyramid was built with thick irregular joints (well-dressed stones only on the exposed face) and gypsum mortar. Only with the Temple at Giza, care was taken to fit the blocks of granite. In the case of Inca construction, it is worth noting how the not aligned courses, both vertical and horizontal, led to walls with appreciable stone interlocking, providing an intrinsic strength against horizontal actions (Peru is one of the countries of South America with the

Figure 3.2 Megalithic irregular masonry. (a) Rubble large stones and (b) polygonal stones.

Figure 3.3 Megalithic walls constructed (a) by Egyptians and (b) by Incas (Peru).

highest seismic hazard). As already pointed out, a similar objective can be achieved through iron connections (e.g. dowels, cramps and dovetails) as shown in Figure 3.1.

In case smaller elements were adopted (e.g. smaller stones or bricks), some ancient builders, especially in Egypt, used to introduce among the courses a bonding or framing timber (or a mat of reeds) to assure integrity and stability of the walls, and also to align masonry courses. Most of the times, the insertion was only along the bed (horizontal courses), but they could also run vertically, creating a sort of net where the masonry represented the infill.

However, it is only with Classical Greek builders that masonry assumed the modern aspect: fine ashlar stones perfectly dressed in parallelepiped shape with no mortar joint. The construction technique maintained the orthogonality of elements in all constructions, either housing, military buildings or temples. As a general remark, the more homogeneity of the masonry and the elimination of the bending stresses on the blocks (thanks to a perfect match of joints) led to a more efficient wall, also with a smaller thickness (Figure 3.4).

The large blocks adopted for the construction of monuments were mainly made of sandstone, as it is more easily handled and prepared. However, starting from the 6th century BC, marble stone started to be imported from Naxos and Paros islands and used in temples. The low tensile strength of stone was an important limitation for Greek architecture, clearly evident in the small distance between the columns of the temples, mostly due to the difficulty of transporting large lintels and the increasing tensile stresses.

A few centuries later, with the aim of achieving higher stability with the minimum economical effort, learning from different civilizations (mainly Hellenistic and Etruscan), Romans started to innovate materials, structural concepts and construction processes. These developments were also due to the vastness of the empire, in constant need of infrastructures, such as roads, bridges and harbours, as well as buildings, temples, houses and aqueducts.

Among the main improvements, Romans invented the concrete (so-called *opus caementicium*) and enhanced the quality of bricks (well-fired), whose size becomes standardized (large and flat) adopting different shapes for different purposes. Moreover, conversely to other previous cultures, Romans were able to differentiate the materials within the building according to the structural functions they were going to perform. They also became great experts of vaulted forms (arches, vaults and domes, discussed in Chapter 4), which allowed them to build constructions over long free spans (e.g. Pantheon in Rome and the Pont du Guard in Nimes).

Considering the walls, the most important innovation regarded the use of three-leaf masonry walls with bricks or small stones as facing, filling the inner space with pozzolanic concrete, which allowed an overall freedom in placing openings. In general, given the high mechanical properties

(a)   (b)

Figure 3.4 Square block masonry. (a) Placed without a pattern and (b) placed with a defined pattern.

of pozzolanic concrete, an adequate connection (interlocking) between interior and exterior leaves was guaranteed simply by the bonding of concrete with the facing, making the wall working as monolithic. Table 3.1 presents the main examples of walls produced by Romans. However, in case the infill was realized only with internal rubble masonry, lacking the internal bond, the wall was composed by three independent leaves, and the connections were essential to avoid leaf separation in the long term (Figure 3.5).

Table 3.1 Examples of walls built by Romans

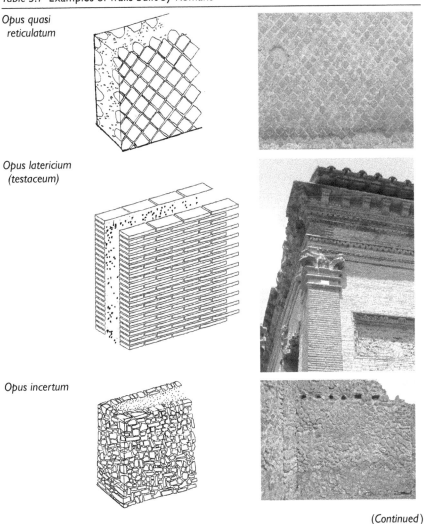

Opus quasi reticulatum	
Opus latericium (testaceum)	
Opus incertum	

(Continued)

Table 3.1 (*Continued*) Examples of walls built by Romans

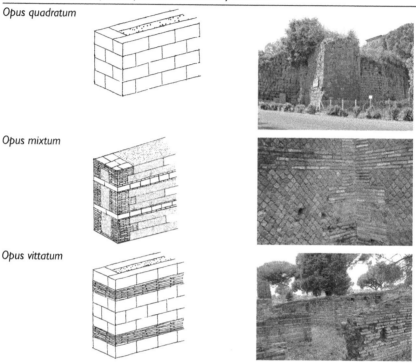

Opus quadratum

Opus mixtum

Opus vittatum

With the fall of the Roman Empire, the Middle Ages represented a period of economic and cultural decline, with an overall deterioration of the construction techniques and materials, pozzolanic concrete in particular. Rather than the lack of secret ingredients, most likely, the quality drop

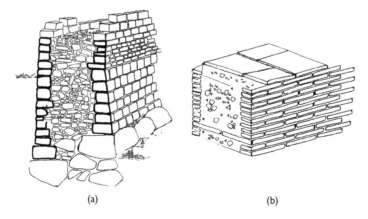

(a)   (b)

Figure 3.5 Three-leaf wall used since the Roman period. (a) Cross section and (b) example of connections between the three leaves.

resulted from having disregarded the basic recommendations to mix pozzolana concrete, as reported in Vitruvius' treatise. For instance, the Roman well-detailed lime kilns were replaced by rudimentary ovens producing uncooked lime; dirty (or mixed with clay) sand was often used, and pozzolana was gradually abandoned, together with the practice of adequately compacting concrete manufactured with low water content. This negligence led inevitably to several failures. Among others, mortar without pozzolanic ash is aerial, taking very long time to harden in thick walls. If it was poorly mixed, possible rain infiltration could leach out the lime, making the infill lose cohesion with the consequent outward (and unsafe) pressure for the external leaves.

During Early Middle Ages, usually referred to as the Romanesque period, the religious orders were pioneers of a new construction phase. One of the most important innovations regarded the wooden roofs: due to the danger of fires, these roofs started to be replaced by masonry arches and barrel vaults. The consequent large lateral thrust led to the construction of thicker walls (and with relatively small openings) or walls stiffened at frequent intervals by flat piers or buttresses to absorb the horizontal action.

Romanesque walls were similar in morphology to the Roman ones (often two external leaves filled with rubble), but the building materials greatly differed across Europe according to local stone and building traditions. In Italy, Poland, much of Germany and the Netherlands, brick was generally used, while other areas (e.g. Portugal) witnessed an extensive use of limestone, granite or flint. The building stone was often used in comparatively small and irregular pieces, laid with thick mortar joints. Smooth ashlar masonry was not a distinguishing feature of the style, particularly in the earlier part of the period, but occurred mainly where easily worked limestone was available.

The subsequent Gothic architecture (starting from the 12th century), integrating extraordinarily aesthetic and structural functions, marked a very important step forward in the field of masonry construction. In this period, the major developments regarded the introduction of arch ribs (for reducing the thickness of the vaults, thus its weight and thrust) and of the pointed arch, leading to an important overall reduction of the horizontal thrust. Aware of the structural function of each element, Gothic builders realized a sort of masonry-framed structures, where the vault thrust was absorbed by flying buttresses and the massive walls were replaced by thinner walls or stained glass openings with no structural function (see Section 3.3.3).

After the Gothic period, limited advances were made in masonry walls until the 19th century, that is, up to the invention of Portland cement. Still, it is noted that this brief review is Europe-centred, with other developments in materials, techniques and architecture (all always interrelated) in different parts of the world, for example India, China or Persia.

For the sake of completeness, in countries in which brick is the primary masonry material, masonry is also distinguished in terms of construction bond (i.e. the geometrical disposition of the bricks along the masonry

Table 3.2 Main brickwork bonds

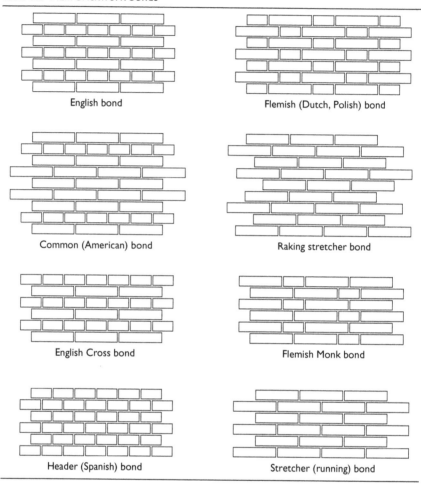

English bond	Flemish (Dutch, Polish) bond
Common (American) bond	Raking stretcher bond
English Cross bond	Flemish Monk bond
Header (Spanish) bond	Stretcher (running) bond

courses), visual aspect, texture and colour combinations between units and mortar. According to the way that the brick is laid and how the brick facing is oriented in the finished wall, different patterns may be formed. Examples are shown in Table 3.2.

3.1.5 Columns

Even if the construction of continuous elements such as walls may be considered natural and spontaneous, isolated structural elements such as columns (used here also indistinctly as pillars or piers) are often needed. The main goal of these vertical elements is to support concentrated loads transferred by arches or beam systems (architraves or lintels), and, contrarily to the walls, they hardly contribute to resist horizontal forces.

One of the most ancient examples of columns is represented by the roofed colonnade entrance to the step Egyptian pyramid at Saqqara. The hall at the end of the passageway was built with 20 pairs of limestone columns with 6.6 m in height, carved to imitate bundled plant stems and built up from superimposed drums. However, the columns were not free-standing, but were attached to the wall by masonry projections. Differently to the subsequent multidrum columns, the earliest monolithic columns (carved from a single rock block) had neither capital nor base. Enlarged ends were probably the consequence of imitation of the first columns made of timber or bundles of reeds. Columns may be tapered with a wider bottom and a thinner top.

As far as the shape of the columns is concerned, spread papyrus, lotus and palm columns, aimed at imitating the form of plants, were typical in the Egyptian architecture (Figure 3.6). These elements were built either with very large blocks (as the Great Colonnade at Luxor) or with small stones. On the other hand, Greek Classical orders (starting from the 8th century BC) deserve a particular mention, as they have represented for centuries the main reference of beauty and elegance all over the world. An order is a certain assemblage of parts subject to uniform established proportions and provided of aesthetic details that make the order readily identifiable. The forms and proportions shown in Figure 3.7 were subsequently assimilated and adapted by Romans. Other famous columns include the ones in the Hall of a Hundred Columns, in Persepolis, Iran, where the capitals adopted the form of twin-headed bulls, eagles or lions, all animals representing royal authority and kingship.

(a)   (b)

Figure 3.6 Egyptian columns resembling plants. (a) Spread papyrus and lotus (in the middle) capitals and (b) palm columns.

Although the Classic orders underwent a revival during the Renaissance, it is during the Middle Ages that the column element was extensively exploited, particularly in the soaring piers of the Gothic cathedrals, becoming a symbol of lightness and grace. However, the extraordinary slenderness was possible only by avoiding lateral actions, even if this meant to overload the elements to get a quasi-vertical force resultant (if necessary by adding weight on top of the columns). This is particularly clear in the Mallorca Cathedral, Spain, where the piers have an octagonal section with a circumscribed diameter of 1.6 or 1.7 m and a height of 22.7 m to the springing of the lateral vaults (Figure 3.8a). The ratio between diameter and height reaches the value of 14.2 (in other Gothic Cathedrals, this value normally stays between 7 and 9), constituting perhaps the more structurally daring aspect of this construction (Roca et al., 2013). A slenderness record seems to be established at Monastery of Jerónimos in Lisbon, Portugal (Figure 3.8b), with columns with about 1 m diameter and 16 m height, achieving a ratio of 15.4 (Lourenço et al., 2007).

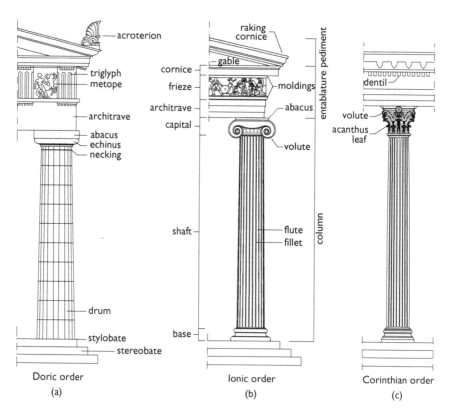

Figure 3.7 Diagram of Greek orders regarding columns and the structural elements above them. (a) Doric order, (b) Ionic order and (c) Corinthian order.

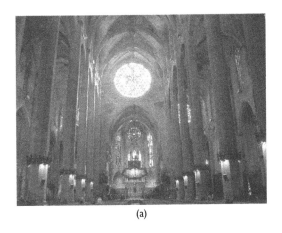

(a)   (b)

Figure 3.8 Column slenderness. (a) Mallorca Cathedral and (b) Jerónimos Church.

From the constructive point of view, during Middle Ages, masonry piers were generally built with an exterior leaf of regular masonry and an interior core of stone or rubble fill. In some cases, they were entirely composed of large blocks adequately interlocked, with no differentiated core. Four different cases are reported in Figure 3.9. In particular, the sonic tomography of one of the Mallorca Cathedral piers shows that it is composed of five stones of similar quality, with a square-shaped central one. The stones rotate 45° in each row to provide satisfactory vertical interlocking. In case the piers were built with bricks, examples of arrangements are reported in Figure 3.10.

3.1.6 Foundations

Compared with the elements already described, foundations do not really represent a further set, being substantially built in a similar way to the previously identified structural components. The importance of foundations for the overall stability of structures is critical and well known from ancient times. The main role of this element is to adequately distribute the load of the superstructure to the ground, guaranteeing acceptable levels of stresses in the soil (or rock). This can be obtained by enlarging the base of the structure in contact with soil foundation (often not the case in historic buildings, as stresses are already distributed in walls), or carrying the loads down to a firmer stratum either by means of a deep substructure or excavation (again often not the case in many historic buildings, which have superficial foundations).

Foundations should also prevent undue movements and excessive deformation, the two most relevant design criterion for modern structures. Like any other structural element, foundations (meaning any soil/structure combination) will experience deformations, including vertical settlement, but also leaning or sliding. In this regard, it is important that the

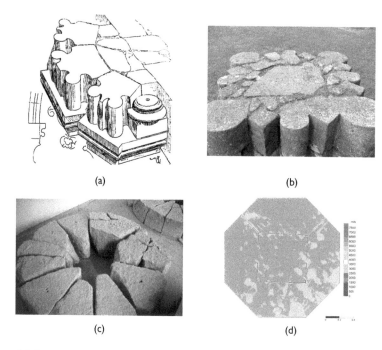

(a)   (b)

(c)   (d)

Figure 3.9 Examples of morphology of Gothic piers. (a) Viollet-le-Duc's drawing, (b) ruins of Arbroath Abbey in Scotland, (c) dismantled pier of Tarazona Cathedral in Spain (core originally filled with poor rubble) and (d) sonic tomography of one of the Mallorca Cathedral piers (large stone blocks).

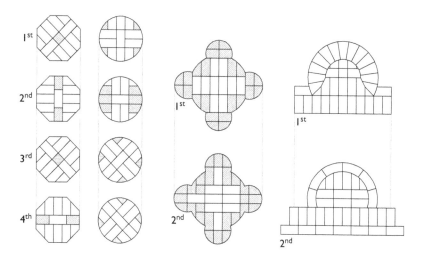

Figure 3.10 Examples of arrangements of brick columns (the ordinal form indicates the course number).

settlement is uniform (within certain limit, also in regard of the level of the surrounding ground) rather than differential. The latter, in fact, can produce more pronounced effects on the superstructure or it can even lead to the tilting of the entire building, as for the Tower of Pisa. In the following, avoiding any discussion on the nature of the ground, which falls out of the scope of this book, attention is paid to the choices of ancient builders to guarantee an adequate support for the building. In any case it is noted that (1) contact pressures in existing buildings are much higher than the usual modern design values; (2) ground under the foundations is often consolidated due to the long-term effects of the gravity loads in the building; and (3) historic buildings have been usually built over long processes, which allowed ancient builders correcting damages and even partly rebuilding elements in case of movements during building erection. Much of the past foundation-induced damage is likely to be stabilized, requiring careful analysis and/or monitoring. Further soil movements, if relevant for the superstructure, are also likely to be a slowly progressive process that may not require immediate action. Finally, it is noted that many constructions are built over archaeological remains and successive layers of the past, meaning that any intervention under the ground must be carefully analysed.

As recommended in ancient treatises, historical foundations were designed according to local experience. Again, it was common practice to use previous existing foundations and remains, as in the case of Gothic cathedrals built over Roman or Romanesque remains, or colonial cathedrals built over pyramids in Mexico. In some cases, the heterogeneity related to combining consolidated and non-consolidated ground created significant problems in the long term. Regarding the soil response, embedding the previous foundations and, again, the overall slow building process contributed to mitigate both the appearance and effect of differential settlements.

From the design point of view, to avoid the effects produced by different soil properties along the structures, ancient treatises recommended to build continuous grid foundations rather than enlarging the bearing surface. Regarding the latter, in case of firm soils, the enlargement of the wall at the foundation could be only one-eighth to one-fourth of its width, depending on the type of wall and the height of the building (it is not uncommon to find no enlargement). Ancient treatises also insisted in the need to ensure the verticality of the axis of the walls/piers and foundations, which has the objective to limit the wall's rotation at the base.

According to local experience, ancient designers were sometimes able to recognize the different capacity of soil layers, and, in many cases, they deepened the foundation until reaching an assumed adequate stratum. This approach is at the basis of the definition (even in modern times) of shallow and deep foundations, where the former consisted of stones (blocks or rubble) and brick masonry (in some cases mixed with reed, small tree branches, cow skins or other additions), whereas the latter could be built, for example, by means of wooden piles. As shown in Figure 3.11a,b, the shallow foundations can be divided into two groups, namely, isolated (for single column footing) and continuous (for wall footing). The two-dimensional extension of the continuous foundation leads to a two-way beam and slab, or flat plate raft.

In turn, the classification of piles can be made with respect to load transmission and functional behaviour. Point-bearing piles transfer their load to a firm stratum, and they derive most of their carrying capacity from the penetration resistance of the soil at the toe of the pile (Figure 3.11c). On the other hand, friction piles derive their capacity mainly from the friction of the soil in contact with the shaft of the pile (Figure 3.11d).

In general, the pile foundation system consisted of a group of driven piles connected by multilayer (mutually orthogonal) battens or planks, or some cap (Figure 3.12a). In case the water table was close to the ground surface (e.g. near rivers, sea or in a lagoon), a higher bearing capacity of the soft superficial ground layers was achieved by shorter and closely spaced piles (to provide confinement and soil consolidation) rather than deep and distant piles. Finally, as a general rule, given the strong susceptibility of timber piles to alternate dry/wet conditions (i.e. variations of the water table), to

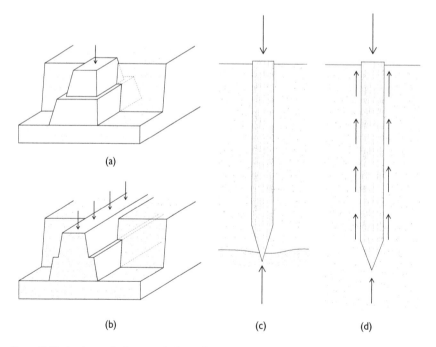

(a)

(b)   (c)   (d)

Figure 3.11 Ancient shallow and deep foundations. (a) Isolated and (b) continuous shallow foundations; (c) point (or end) bearing and (d) friction piles (deep foundations).

avoid decay and rot, timber elements were ideally located above or under the water table. It was well known that timber piles submerged in ground water tend to last indefinitely.

The operations needed to drive the piles are described in several treatises of the past, as Francesco di Giorgio Martini's. Generally, a heavy weight (e.g. stone block) was raised, thanks to a mechanical device, and then released to hit the pile. To protect and strengthen the tip of the piles, these were often provided with a steel toe cover, facilitating also the penetration of the pile into the ground (Figure 3.12a). On the other hand, the execution could damage the timber piles, causing the deterioration of the tip or the rupture of the shaft due to incorrect driving procedures or firmer strata encountered (Figure 3.12b).

For the sake of completeness, some comments related to the analysis of foundations in compliance with current standards are stressed. According to modern engineering, the soil-bearing capacity is evaluated as the maximum pressure that can be achieved at the subgrade, without causing large settlements, which would damage the structure. Although this criterion is acceptable for new foundations, it should not be directly applied to problems concerning ancient constructions (see Chapter 1). First, giving the lasting construction process and the nature of timber and masonry itself, ancient structures were able to accommodate large settlements or deformations with little or moderate damage. Second, the settlement may have been already stabilized centuries or decades ago, which means that the process is mostly over, and no further damages will be experienced by the construction. Finally, damage caused by settlements is often easily repairable, for example settlement cracks along the joints could be just refilled, if they do not significantly affect the structural safety.

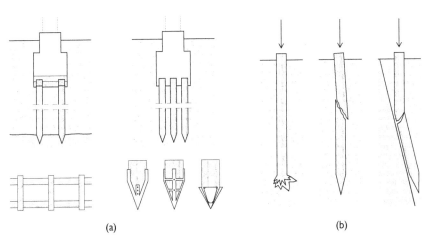

Figure 3.12 Ancient deep foundations on driven piles. (a) Arrangements and (b) related problems.

3.2 TIMBER

Timber, together with stone, is one of the oldest building materials used by mankind. Its importance relies on the fact that it allowed men to settle in regions where the climate was very different from Africa, for example colder Mediterranean and continental areas. Besides that, timber was the first material adopted in ancient times with good tensile strength. This made it, until the 18th century (i.e. until iron started to be widely used), the best material adopted for building horizontal elements (subjected to bending), such as beams and floors. Among structural materials, timber is the only one that, directly available in nature, undergoes structural loading comparable with the one experienced in its natural environment.

3.2.1 Wooden buildings in the antiquity

The origins of timber structures are lost in the mists of time. According to the archaeological evidences found in the major European archaeological sites (e.g. Bilzingsleben, Port Pignot, La Roche Gélétan, Grotte du Lazaret), from 320000 to 120000 BC, wooden poles or branches were probably used as bearing structures of rudimentary and temporary tents used as shelters for hunters (Figure 3.13). Wooden fragments of built structures cut by a primitive saw were dug out in Věstonice in Moravia (30000–25000 BC), whereas a permanent settlement with rectangular dwellings made from squared timber was found in Monte Verde in Chile (around 15000 BC).

With the advent of masonry, the first mixed masonry-timber structures started to be built. One of the earliest buildings made by stone and clay walls reinforced by vertical wooden poles was found at Beidha in Jordan (8000 BC), while the first timber roof, 3.8 m long, was found in the Yokoo site near Oita in Japan (8000–4000 BC). It is in this period (Neolithic) that timber houses started to appear in Europe with an advanced post-and-lintel system (Figure 3.14). Figure 3.15a shows the reconstruction of a European early Neolithic long house (5000–4000 BC), predecessor of timber framing systems. The remains of a well found in Saxony and built with perforated

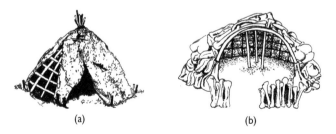

(a)   (b)

Figure 3.13 Reconstruction of earlier Ukrainian shelters. (a) Low domical shape covered with animal skin and (b) tent restrained by heaped mammoth bones and tusks (predecessor of the cruck-frame bearing system).

wooden boards are a remarkable example of the high developed carpentry of that age (Figure 3.15b).

If timber was abundant in Europe, Northern Africa was lacking large forests and tall trees. In Egypt, for instance, the native timber was mostly inadequate for structural applications and could only be cut into short planks. Acacia, carob, juniper, palm, sycamore and some other local woods were used, while hardwoods like ebony were imported from eastern Africa, and structural cedar and pine were imported from Lebanon.

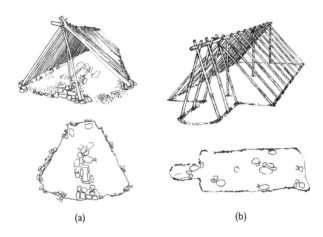

(a)   (b)

Figure 3.14 Reconstructions of the earliest timber structures based on excavated sites (predecessor of the king post and ridge systems). (a) Lepenski Vir, Serbia, 5000 BC (by the Danube River) and (b) Vuollerim, Sweden, 4000 BC (beyond the Polar Circle).

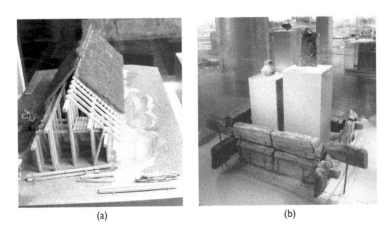

(a)   (b)

Figure 3.15 Neolithic European carpentry skills. (a) Model of early long house, predecessor of aisled timber framing systems (5000, 4000 BC) and (b) remains of a Neolithic well found in Saxony.

As seen in Section 3.1.5, stone elements were sometimes influenced by tectonic forms of former timber or cane structures. This process affected also the developments of Greek architecture, particularly of temples. The Early Archaic temples were built with wooden superstructure (columns and entablature) and the existing adobe walls were often reinforced by wooden posts, similar to half-timbered technique. Subsequently, the elements of this simple and clearly structured wooden architecture were extended to stone temples. This is especially evident in the friezes, where the triglyphs are thought to be a representation in stone of the former timber joist ends of early temples (Figures 3.7 and 3.16).

In Asia, there are some of the oldest still existent timber constructions. The core temple of Horyu-ji in Nara (Japan), founded in 607 AD, is known as the oldest surviving wooden structure in the world, and the Goju-no-tou (a five-storey Pagoda), standing to the west of the Kondo (main building), being 32.6 m high is the one of the oldest wooden towers in the world (see Figure 3.17). According to a dendrochronological analysis, its central pillar dates back to 594 AD (dendrochronology is the science or technique of dating events, environmental change and archaeological artefacts by using the characteristic patterns of annual growth rings in timber and tree trunks).

3.2.2 Columns, beams and floors

The use of wooden elements to span between two supports and to create floors or roofs is an ancient procedure that probably followed from the observation of a fallen log bridging a gap. In ancient times, the first floors probably consisted of side-by-side logs covered with turf, reed matting, earth or other materials, whereas some inclination was necessary for building roofs. On the other hand, the first column presumably consisted of branches or logs simply fixed into the ground to provide them some

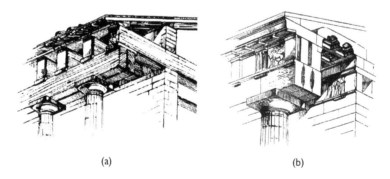

(a)   (b)

Figure 3.16 Triglyph development in ancient Greek temples. (a) Early wooden temples and (b) subsequently stone temple.

(a)   (b)

Figure 3.17 Horyu-ji in Nara (Japan). (a) Possibly the oldest wooden structure in the world and (b) the five-storey pagoda.

resistance to overturning moments. This technique makes the buried part of the log continuously exposed to the dampness of the ground, which usually brings wood to a fast decay.

Although stone was preferred in comparison to wood for building vertical elements, due to its higher stability, stiffness and bearing capacity, wood was used for millennia to cover spaces in the forms of beams and planks. The main drawback of timber was the limitation in the span, related to the available length, transportation and placement, and the strength, probably the cause of many long narrow rooms of the past. In order to cover larger spaces, ancient population widely exploited the famous cedar of Lebanon (e.g. Egyptians, Persians or Romans).

This geometric limitation, intimately connected to the use of readily available timber (the only ancient process was the careful shaping of the trunk in a square or rectangular shape), was partially overcome at the end of the Roman Empire with the development of truss construction (see also Section 3.2.3). Unfortunately, being wooden structures prone to decay and susceptible to fire, many examples of the past are lost, and the technical solutions developed by Romans and during Middle Ages are only described in treatises. As shown in Figure 3.18, ancient builders achieved larger spans either by ingenious two-dimensional assemblages of short elements (i.e. grids, thanks to the bending moment capacity of wood elements) or by unidimensional longer beams built up with shorter elements (thanks to more or less complicated joinery) (see Mainstone, 1998). In both cases, the capacity of the floor or of the beam was almost exclusively determined by the connections.

Apparently, no significant evolution in the construction techniques were notable until the invention of glued laminated timber (starting around 1850), which allows to build elements of any shape and size (see Sections 3.2.6 and 4.3.3).

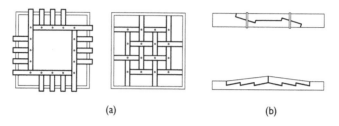

(a)   (b)

Figure 3.18 Examples of early timber complex developments. (a) Floors (Villard de Honnecourt, 1230; Serlio, 1619) and (b) beams.

3.2.3 Roof trusses

Although the construction of columns, beams and floors may have followed natural prototypes, the construction of trusses, as assemblages of smaller elements (acting as struts and ties in modern terms), probably represents the final stage of a continuous research performed by ancient builders (by trial and error). There is no doubt that the most important and difficult role was played by the joints, as discussed in Section 3.2.5.

The difficulties posed by bending moments at connections probably led ancient builders to build trusses in a triangular form. In this case, stability is guaranteed by the shape itself, being the triangle the simplest geometric figure that does not change shape when the lengths of the sides are fixed. In structural terms, a triangular truss conveys the loads to the supports with no need of transmitting bending moments, meaning that the individual elements may experience a relative rotation at the connections without compromising the stability of the structure.

In Greek classical architecture, there is limited evidence of the adoption of timber trusses for covering wide spanning spaces, even if this is plausible. The awareness of triangular forms may be seen in the Lion Gate (Figure 3.19), erected during the 13th century BC in Mycenae (southern Greece). In this analogy, the lions represent the two rafters, whereas the central pillar is the post (nomenclature is addressed later). In general, it is not clear if Greeks were aware of the basic concept of the form (with the post not connected to the tie beam) or if they used a truss on several supports.

Similarly, the only written evidence of the Roman truss is reported in Vitruvius' treatise, but his description is rather vague, simply referring to a system of two inclined rafters connected by a horizontal tie beam. The 16th-century fresco by Giovanni Battista Ricci (nicknamed *Il Novara* after his birth town), together with the drawing of 1693 by Giovanni Ciampini, represent an important evidence of the roof truss of the 4th-century Old Saint Peter's Basilica (Figure 3.20). The two pictures display a clear triangular shape truss, where double tie-beam system and the absence of inclined struts reveal some lack of understanding in the basic concepts of the form.

Figure 3.19 Lion Gate in Mycenae, southern Greece (13th century BC).

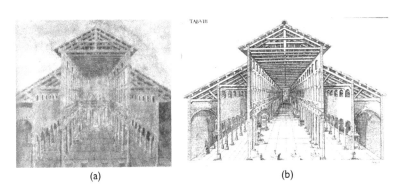

(a)   (b)

Figure 3.20 Fourth-century Old Saint Peter's Basilica. (a) Fresco by Giovanni Battista
Ricci located in the sacristy of the basilica (16th century) and (b) drawing by
Giovanni Ciampini (1693).

The 13 roof trusses of Saint Catherine's Monastery on Mount Sinai may
be the oldest roof trusses in the world (Figure 3.21). Build by the architect
Stephen of Aila during the 6th century AD, this type of truss may have been
the same adopted for centuries by Romans and in Paleo-Christian basilicas (where vaults were not adopted). Together with the triangular truss, a
secondary system made by a central post and two inclined struts can be
recognized. The post is not connected to the horizontal tie beam, but it
actually works as a tie for the two inclined struts. In turn, the two short
struts represent an intermediate support for the rafters against sagging.

Although, in southern Europe, the ancient builders kept building roof
trusses in the fashion of the one in Saint Catherine's Monastery, this shape

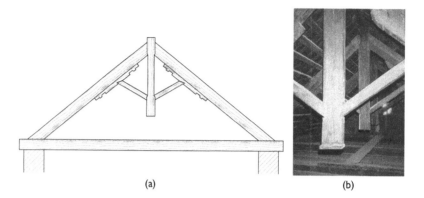

(a)   (b)

Figure 3.21 Roof structure of the nave for the Church of Saint Catherine, Mount Sinai. (a) Drawing and (b) detail of the king post and inclined struts.

was not suitable for the steeper roofs of northwest Europe. Moreover, in many cases, the presence of vaults or cupolas would not allow to include tie beams at the support level. In such case, the roof truss was either supported by the vault itself or composed of two simple rafters connected by few struts to prevent sagging (Figure 3.22a). This type of roof is illustrated in the drawing of Villard de Honnecourt, where the wall plates (the horizontal timber elements along the top of the masonry walls) were placed below the level of the vault crown (Figure 3.22b). In order to reduce the horizontal thrusting, a later modification was represented by the arch-braced truss (Figure 3.22c). In other cases, when a continuous tie beam was necessary, the entire roof truss was placed above the level of the vault crown, as in Notre Dame in Paris (Figure 3.22d).

An interesting innovation was the hammerbeam truss (made of oak wood) of the Westminster Hall in London, built in 1399, covering a space 20.8 m wide and 73 m long (Figure 3.23). This span, resulting from the

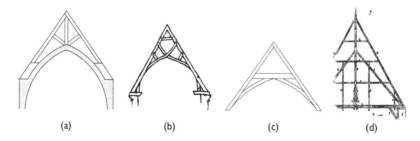

(a)   (b)   (c)   (d)

Figure 3.22 Examples of steep roofs. (a) Truss supported by the vault; (b) vaults with no tie beams at the wall plates level, by Villard de Honnecourt (1230); (c) arch-braced truss; and (d) one of the trusses above the vaults of Notre Dame in Paris.

Figure 3.23 Westminster Hall in London (1399): interior view by Thomas Rowlandson and Augustus Pugin.

removal of two rows of columns, which originally divided the hall into a central nave and lateral aisles, almost approached the one of Roman and Paleo-Christian buildings. Despite the extensive use of triangular forms, according to the actual level of rigidity of the connections, it is not easy to discuss how this truss works. Its shape suggests an accurate assemblage of former types of truss, where the arch represents the main supporting structure.

In Italy, the shape of the roof truss of Saint Catherine's Monastery was consolidated, and during the Renaissance, many elaborations on this solution started to be implemented. The solution proposed by Palladio for covering the Teatro Olimpico in Vicenza, built at the end of the 16th century, is one of the most brilliant examples (Figure 3.24a). The Pantheon roof is another example of Palladio's work, although made of bronze. It was removed by Pope Urban VIII in the early 17th century to make guns and cannons for the rampart of Castel Sant'Angelo in Rome. Despite the fact that the truss was basically composed of triangular forms, it had two intermediate supports and lacked a clear tie beam (i.e. it did not prevent the rafters from thrusting outside). All the elements seem to have been fabricated from plates of bronze and riveted together at the intersections. Being probably one of the first structures of this kind realized in metal, apparently no differences between strut and tie elements were implemented to prevent buckling.

Other Italian truss configurations were proposed by Serlio in his treatise of 1584 and are shown in Figure 3.24b.

One of the last structural topologies of roof trusses is represented by the scissor truss shown in Figure 3.25, conceived to cover medium-span rooms. The main advantage with respect to the former shapes depicted in Figure 3.22b relies on the usage of straight elements.

A multiplication of forms started to appear all over Europe during the Renaissance and Baroque periods. An impressive example of later roof truss is represented by the Exerzierhaus, built in 1771 in Darmstadt (Germany). Covering a free span of 45 m, it is the largest roof of pre-industrial times (Figure 3.26).

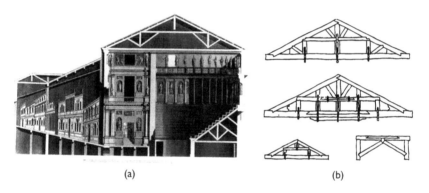

(a)   (b)

Figure 3.24 Examples of Italian Renaissance roof trusses. (a) Palladio's Teatro Olimpico in Vicenza (16th-century, drawing by Ottavio Bertotti Scamozzi, 1776) and (b) roof truss drawings by Sebastiano Serlio (1584).

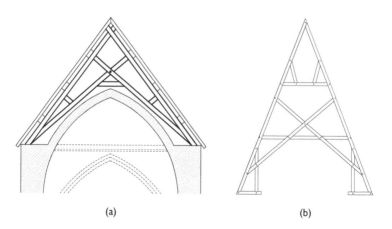

(a)   (b)

Figure 3.25 Examples of scissor truss. (a) Front view (possibility of accommodating the upper part of vaults or curved ceilings) and (b) elaborated version.

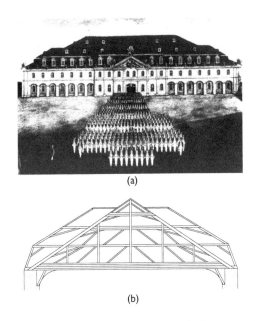

(a)

(b)

Figure 3.26 The largest roof of pre-industrial times. (a) Elevation of the building and (b) cross section with truss.

Finally, for the sake of clarity, the subsequent figures describe the main types of roof trusses, defining the designations of the truss itself, see Figure 3.27, and of its constitutive elements, see Figures 3.28–3.31.

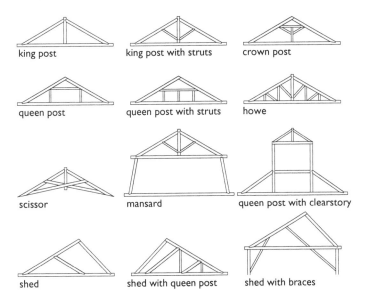

Figure 3.27 Common timber truss configurations.

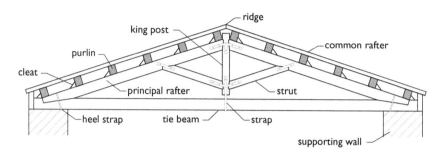

Figure 3.28 Constitutive elements for a king post truss.

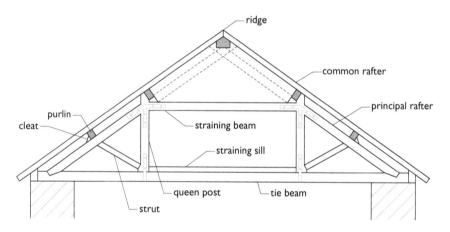

Figure 3.29 Constitutive elements for a queen post truss.

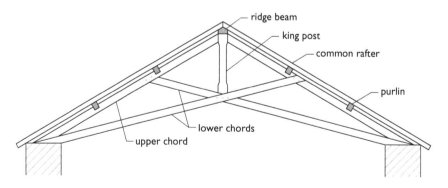

Figure 3.30 Constitutive elements for a scissor truss.

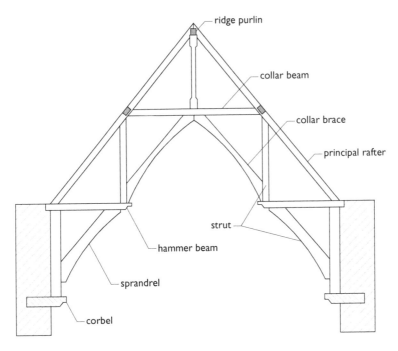

Figure 3.31 Constitutive elements for a hammerbeam truss.

3.2.4 Bridges

Compared with their predecessors, following the techniques developed by Etruscans, Romans made major advances in timber technology, especially in the field of military engineering, with direct consequences on bridges. Built in only 10 days, between 55 and 53 BC (and destroyed soon after the exploration of the northern regions), the first of the two Caesar's Rhine bridges was an outstanding example of Roman skills, symbol itself of the expansion of the empire (Figure 3.32a). Another important bridge was built over the Danube, in Romania, by Apollodorus of Damascus in 103–105 AD (also depicted in the reliefs on Trajan's Column). The truss, a hollowed-out beam with the forces concentrated in a triangulated network of linear members, was apparently a Roman invention (Figure 3.32b).

After the fall of the Roman Empire, no significant contributions seem to have been made until Renaissance. Subsequent to the works of Villard de Honnecourt and Leonardo da Vinci, fully detailed drawings for bridge trusses are reported in Palladio's treatise of 1570. Figure 3.33a shows the Alpini's bridge (also referred to as Old Bridge) designed in 1569. It is a covered wooden bridge located in Bassano del Grappa over the Brenta River, in Italy. It was destroyed and reconstructed many times, the last of which

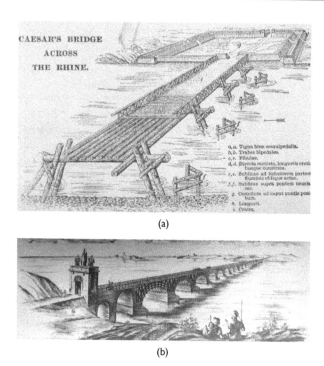

Figure 3.32 Reconstruction of two Roman Empire bridges. (a) Caesar's Rhine bridges built in 10 days between 55 and 53 BC and (b) bridge over the Danube built in 103–105 AD.

in the Second World War. Figure 3.33b shows the project of the bridge over Cismon River (with a span of around 34 m), while Figure 3.33c–e provides the proposals for three bridges without intermediate supports.

In the following centuries, many timber road bridges were built, but it is during the 18th century that, thanks to science improvements and the study of catenary, a new boost was given to bridge design, culminating with the Grubenmann brothers' bridges. As an example of these, the Wettingen Bridge in Switzerland, built in 1765, with a free span of 62 m is the largest bridge of pre-industrial times (Figure 3.34). The bridge combines the arch and truss principle with seven oak beams bound together to form a catenary arch to which a timber truss is fixed.

Finally, during the 19th century, the continuous demand for roads and railway bridges in North America led to the final stage of beam-like truss evolution, achieving the modern shape currently used, as discussed in Section 4.3.3, Glulam and Trusses. An important awareness was the distinction between struts and ties for the diagonal elements, with the progressive replacement of the ties by iron. This period also marked the transition from timber to iron bridges.

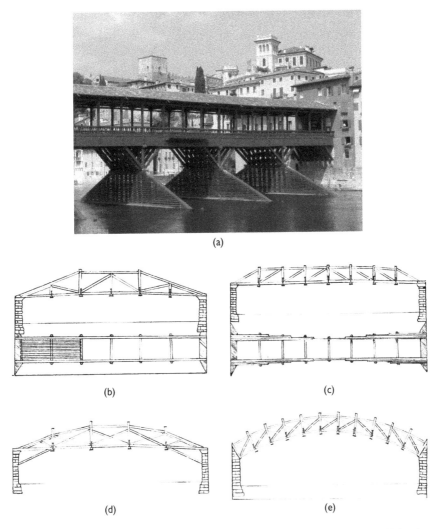

Figure 3.33 Palladio's bridges. (a) Alpini's Bridge in Bassano del Grappa, Italy; (b) bridge over the Cismone River (elevation and plan); and (c)–(e) three alternative designs for bridges without intermediated supports.

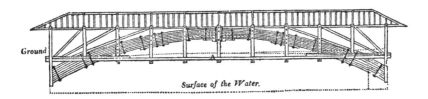

Figure 3.34 Wettingen Bridge in Switzerland built in 1765.

3.2.5 Other timber structures and techniques

As stated earlier, the origin of timber structures dates back to early man's dwellings and shelters. Starting with huts and stilt houses (raised on piles over the surface of the soil or a body of water), as the centuries pass, techniques and applications of timber have been continuously refined to achieve a more efficient use. Where local forests provided a convenient supply of timber as a building material (e.g. in North Europe), ancient builders realized constructions entirely made of logs. In most cases, timber framing and 'post-and-beam' systems were introduced for building structures using carefully fitted and joined timbers elements. The frame system has been used for thousands of years in many parts of the world, but it is considered typical of Central European buildings realized from the Middle Ages to the 19th century. Also in Asia, this technique has been widely adopted, most of the time, for building temples. From the structural point of view, the basic idea is that a timber frame can resist tension, thus providing a better resistance to horizontal loads. Moreover, the frame provides confinement for the infill (especially, masonry panels), improving their mechanical properties against shear loads.

Since architecture and techniques have evolved over the centuries according to local traditions, nowadays, the styles of historic framing are categorized by the type of foundation, walls, how and where the beams intersect the supporting elements, the use of curved timber elements and the roof framing details. In the following, a brief description of the most common techniques used for different purposes is given together with illustrative pictures.

3.2.5.1 Log house

In places where timber of large enough diameter was available, logs were simply placed on top of each other (in the same way as masonry) and notched at each end to interlock perpendicular walls and achieve the requested stability (Figure 3.35). This was the most common building technique in large regions of North Europe and America, where straight and tall trees were readily available (e.g. coniferous, whose resin also contributes to protect the wood). The log house was also a widely used technique for vernacular buildings in the cold regions of Eastern Central Europe and parts of Asia. In warmer countries, where deciduous trees predominate (i.e. trees that seasonally shed leaves) timber framing was favoured instead.

3.2.5.2 Cruck frame

The cruck frame is realized with long, generally naturally curved, timber members that lean inwards and form the ridge of the roof. In order to have a symmetric frame, the original timber member was commonly sliced longways down the middle so that, whatever the shape of the curve, the two sides (i.e. cruck blades) were symmetrical. The two blades were generally secured by a horizontal tie beam (or 'collar') that forms an 'A' shape

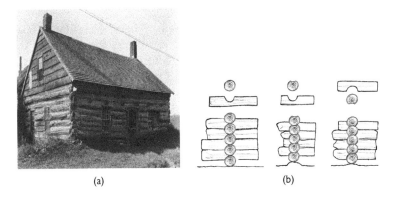

(a) (b)

Figure 3.35 Log house. (a) External view and (b) early log house corner details.

(Figure 3.36). Two or more frames were built on the ground and then lifted into position, then connected together by solid walls or cross beams to provide stability to the construction (i.e. to prevent racking, the action of each individual frame going out of square with the rest of the frame, and thus risking collapse).

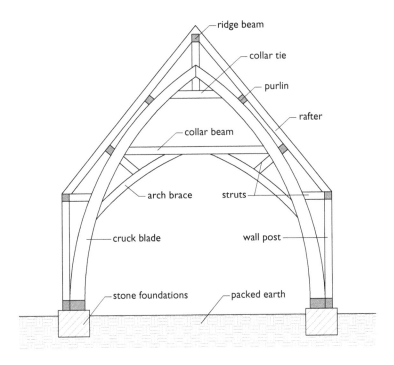

Figure 3.36 Cruck frame details.

3.2.5.3 Half-timbered construction

Despite the ancient use of timber elements to reinforce masonry construction, timber framing, as currently known, was probably introduced during the Roman Empire, for example, *Casa a Graticcio* found in the archaeological site of Ercolano (buried by the eruption of Vesuvius in 79 AD), and usually referred to as *Opus Craticium*. Half-timbered constructions were used not only throughout Europe, such as Portugal (*edifícios pombalinos*), Italy (*casa baraccata*), Germany (*fachwerk*), Greece, France (*colombages* or *pan de bois*), Scandinavia, United Kingdom (*half-timber*), and Spain (*entramados*), but also in India (*dhaji-dewari*), Turkey (*himis*) and other locations. The technique allows combining poor materials (rubble masonry and small timber elements) to obtain structural walls, usually much thinner than masonry walls.

Nowadays, half-timbering refers to a structure with a frame of load-bearing timber, creating spaces between the timber elements (panels), which are then filled in with non-structural material (infill), for example earth, stone, bricks or any rubble. The frame is composed of vertical, horizontal and diagonal elements, being the last one of capital importance for the stability of construction, even if not always present. The exterior timber frame was often left exposed, providing a unique aesthetics to this kind of buildings (Figure 3.37).

Figure 3.38 shows an example of technology transfer from timber structures to other materials. The Tower of All Saints' Church in Earls Barton, England, was built with stone rubble and was decorated in the façade with limestone slabs and strapwork, resembling the aspect of half-timbered constructions.

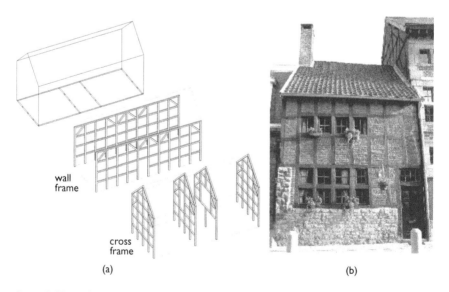

Figure 3.37 Half timber. (a) Different components of the system and (b) house in Theux, Belgium, 17th–18th centuries.

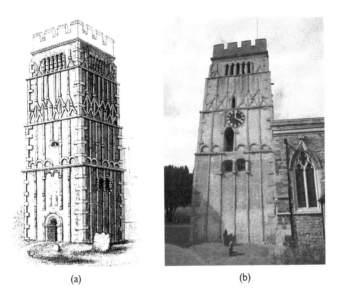

(a)   (b)

Figure 3.38 Inspiration from timber structures to other materials: Tower of All Saints' Church in Earls Barton (England), 10th century. (a) Engraving image with the later medieval battlements (the Saxon roof may have been pyramidal) and (b) present aspect.

3.2.5.4 Jettying

Jettying is a building technique used in medieval timber-frame buildings, in which an upper floor extends beyond the dimensions of the floor below, see Figure 3.39. This has the advantage of increasing the available space in the building without obstructing the lower part of the street, as required by the applicable regulation.

3.2.5.5 Thatched roofs

This type of roof consists in dry vegetation, such as straw, water reed, sedge, rushes or heather, layered in such a way to shed water away from the inner roof. Thatching represents a very old roofing technique and has been used above all in North Europe but also in countries with both tropical and temperate climates (e.g. Kenya and Japan), see Figure 3.40.

3.2.5.6 Pagodas

A pagoda is a tiered tower with multiple eaves, typically curved upward, see Figure 3.41. Pagodas were traditionally built in Buddhist temple complexes, usually found in East and Southeast Asia, where this was long the prevailing religion. Despite the significant resistance to earthquakes, along the centuries, many Pagodas have burnt down, due to fire. Moreover,

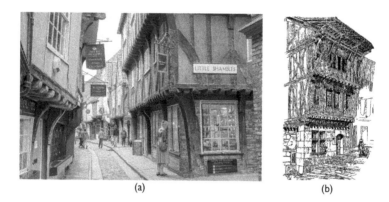

(a)   (b)

Figure 3.39 Jettying technique. (a) Example in a historical centre and (b) detail of timber structure.

(a)   (b)

Figure 3.40 Thatched roofs with wooden structure. (a) Construction process and (b) example in Madeira island, Portugal.

(a)   (b)

Figure 3.41 Examples of Pagodas in China. (a) Chengling Pagoda of Zhengding (1161–1189 AD) and (b) the Khitan wooden Yingxian Pagoda (1056 AD).

because of their height and the possible presence of a metal decorative ornament (finial) at the top, pagodas have attracted lightning strikes and have suffered related damage.

3.2.5.7 Masonry buildings' tying

Regarding masonry construction, timber floors or roofs often played an important role in tying opposite or orthogonal walls, see Figure 3.42, improving significantly their out-of-plane behaviour. This feature is particularly helpful in case of seismic event as it guarantees the box (or integral) behaviour of the building.

3.2.5.8 Centring

Centring is a temporary timberwork used by stonemasons to achieve the correct geometry of vaulted system and roof spires (Figure 3.43). The timber structure was also needed as scaffolding (i.e. temporary support).

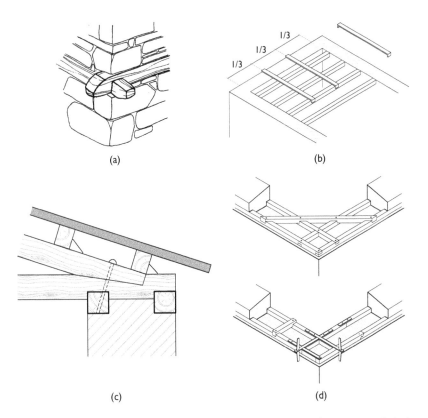

(a)

(b)

(c)

(d)

Figure 3.42 Examples of connection of masonry structures by means of timber elements when adequately detailed. (a) External bracing, (b) timber floor, (c) roof and (d) corner keys.

Figure 3.43 Examples of centring. (a) Roman arch; (b) Walnut Lane Bridge, USA (1908); church of Spisska Kapitula, Cathedral of Saint Martin, in Slovakia (16th century): (c) external photo and (d) internal view of the northern spire.

3.2.6 Woodworking and historical joints

3.2.6.1 Woodworking

Woodworking represents one of mankind's first skills with the consequent development of tools to shape wood (Figure 3.44). With no doubts, the ability improved with technological advances, particularly related to metalworking. It is in ancient Egypt that, starting from 3100 BC, new tools started to be invented and developed, paving the way to a variety of timber joints. This is the case of the original copper tools, then made of bronze, as the ones shown in Figure 3.45a (chisels, axes, adzes, pull saws, etc.). Egyptians introduced also the use of the bow drill, whose application is illustrated in Figure 3.45b.

Remarkable are also the Chinese improvements in carpentry during their Spring and Autumn period (771–476 BC), when intricate glueless and nailless joinery started to be developed, achieving much recognition. At the time, among others, plane and chalk-line tools were introduced, see Figure 3.46.

In Europe, after the advances of the Roman Empire, no significant changes were made during the following centuries, being the medieval tools (Figure 3.47) quite similar to the modern ones (except for the use of steel instead of iron for cutting edges).

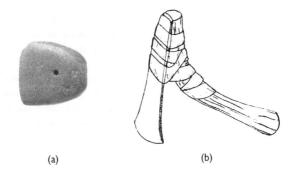

(a)   (b)

Figure 3.44 Ancient woodworking tools. (a) 6,000-year-old stone axe blade and (b) drawing of traditional adze cutting tool.

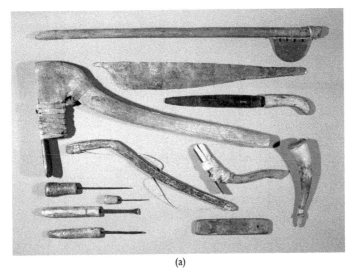

(a)

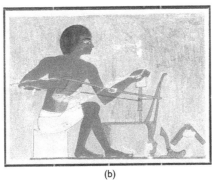

(b)

Figure 3.45 Egyptian woodworking tools. (a) Group of bronze tools found in Thebes (around 1300 BC) and (b) painting regarding the use of the bow drill.

Figure 3.46 Examples of tools in carpentry. (a) Plane used to flatten, reduce the thickness of and impart a smooth surface to a rough piece of timber and (b) chalk line for marking long, straight lines for working timber in a rough and non-planed state.

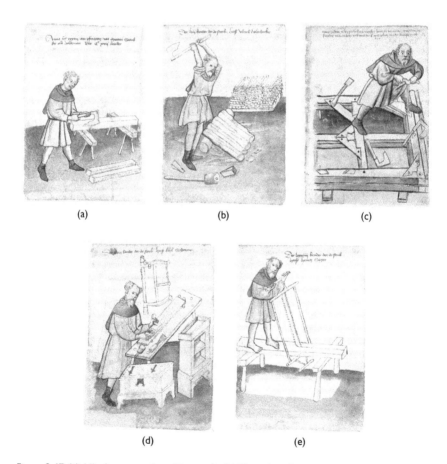

Figure 3.47 Middle Ages woodworking tools. (a) Short-handled broad axe; (b) long-handled bearded axe; (c) auger (15th century); (d) traditional plane and bow saw; and (e) rip sawing with a frame (or sash) saw on trestles rather than over a saw pit.

3.2.6.2 Historical joints

Joints are special zones or devices allowing the structural connection of two or more structural members to produce a more complex structure. The critical stresses experienced in the joints are normally direct tension with shear or are associated with bending. A first distinction can be done between connections of timber elements and connections of timber elements and other bearing systems, for example masonry walls or columns (Tampone, 1996). Although the connections of the second typology are simpler because timber members usually rest on the bearing elements (sometimes with the help of extra elements to distribute the stresses or to protect wood from moisture), along the history many possible timber-to-timber joints were developed. Table 3.3 presents a schematic synthesis of the main historical joints in timber structures (i.e. timber-to-timber).

According to Table 3.3, the connections of the two first categories work mostly, thanks to a combination of simple notches, correct placement of timber elements and geometric detailing. The transmission of the stresses is, thus, based on the contact (including shear) and the contact pressure induced by the gravitational loads on the elements, in some cases, enforced by wooden pegs and wooden wedges.

Lashing (i.e. ropework) represents probably the first type of timber joints, widely used in the prehistory, but also in ancient Egypt (Figure 3.48). In Egypt, in particular, timber joints have been highly developed, starting with leather or cord lashings, but resorting also to wrought iron nails, pegs and dowels, and even animal glue (adhesive created by prolonged boiling of animal connective tissue) in a later stage (after 2000 BC). This advanced technique was probably developed first for joinery (i.e. the skill of fitting pieces of wood together precisely) of handcrafted furniture and then extended to construction.

One of the most used timber connections is the so-called *mortise and tenon*, whose variations have been the basis of most timber joints. According to Figure 3.49, the mortise is the slot or cavity cut into the wood, while the tenon is the projection end of a timber element that is inserted into the mortise. Wooden pegs (and later nails) can be used, but are not present in many cases.

Table 3.3 Classification of the main historical joints in timber structures

Primitive joints Easy-to-use	Sophisticated joints (timber framing) Special tools required	Reinforced joints (shear stress resistance) With extra components
Lashing (roping)	Overlapping	Dowels
Tongue and fork (natural)	Mortise and tenon	Glue
	Housing	Metals (nailing)
	Bird's mouth	

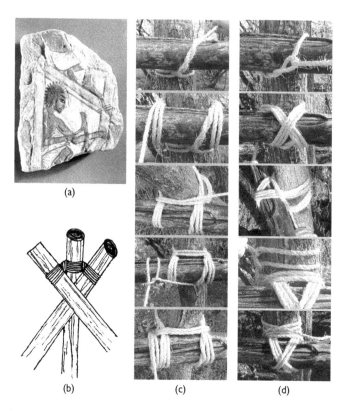

(a)

(b)   (c)   (d)

Figure 3.48 Timber connection by lashing (convenient for round poles connection). (a) Drawing of ancient Egypt, (b) schematic representation of a tripod lashing, (c) square and (d) diagonal lashing.

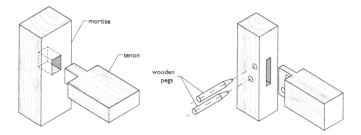

Figure 3.49 Schematic representations of mortise and tenon joint.

Romans were skilled experts in shaping timber elements and connecting them. Their main connection types are shown in Figure 3.50. No significant differences were noticeable during the following centuries, except for the use of metallic connections, for example nails, screws, anchors and dowels, increasing significantly the shear and bending resistance of the

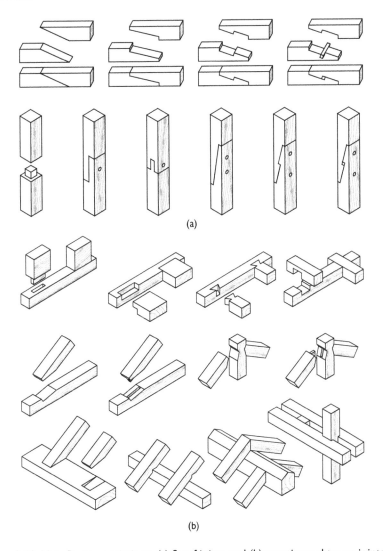

Figure 3.50 Main Roman variations. (a) Scarf joints and (b) mortise and tenon joints.

joint, thus allowing some transmission of bending moment between the members of a frame.

Chinese carpentry is commonly referred to as the most advanced and refined in timber joints, as stated earlier, distinguished by intricate glueless and nailless connections (Figure 3.51a). Chinese have invented also a wide range of fasteners to hold pieces of wood together, exporting their skills to other Asian countries, for example India, Japan and Korea. One of the most important elements in traditional Chinese architecture is represented by the dougong brackets, which are a unique part for interlocking wooden

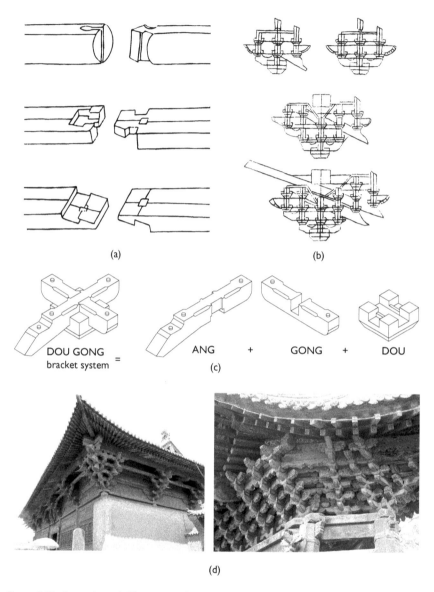

(a)

(b)

DOU GONG
bracket system =

ANG   +   GONG   +   DOU

(c)

(d)

Figure 3.51 Examples of Chinese timber joints. (a) Forms of tenon and mortise, lap and scarf joints, (b) dougong brackets from building manual 'Yingzao Fashi' by Li Jie (1103), (c) schematic representation and (d) images of this type of connection.

brackets (Figure 3.51b,c). Figure 3.51d shows the impressive aspect of these connections in real buildings.

Along the history of timber construction, the main difficulty has always been the detailing of joints able to adequately transfer tensile stresses.

This issue inevitably reflected itself in the limited dimensions (length, particularly) of timber elements, in some cases, coinciding with the size of the trunk. Starting from the Renaissance, different solutions for building long beams have been proposed (Figure 3.52).

A significant improvement has been made in the second half of the 19th century when, thanks to new synthetic glues and cutting tools, it was possible to build laminated timber beams. Today, glued laminated timber is a structural element composed of layers of small thickness of timber finger-jointed and bonded together with moisture-resistant structural adhesives. As long as the joints are well lapped, the beam can be of any length and shape (e.g. even curved and twisted). See Section 4.3.3 for further details.

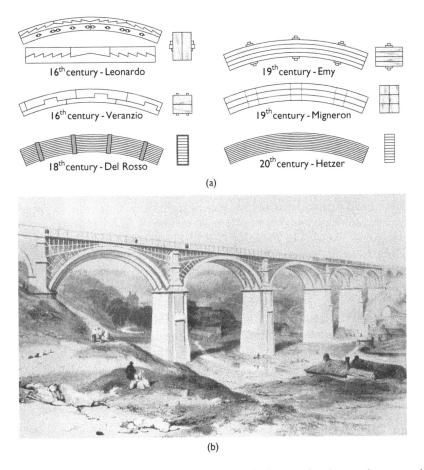

Figure 3.52 Glue laminated timber. (a) Historical designs for longer beams and (b) example of a large-span structure (Ouseburn bridge, built by the Newcastle and North Shields Railway in the 1830s).

3.3 METALS

In ancient times, copper, bronze, iron and lead were mostly used for ancillary building components rather than elements with structural purposes. For structural purpose, iron was widely implemented only after the industrial revolution. However, as stressed in the previous sections, the development of metal tools accompanied the evolution of timber and stone, thus of the overall construction history. In order to have a timeline, the history of metals is usually divided into copper age (from 6000 BC), bronze age (from 3000 BC) and iron age (from 1000 BC). Copper and bronze have been used for roof cladding since Egyptian times.

An important breakthrough was the introduction of wrought iron (very low carbon) or 'worked' iron, whose know-how was brought by Hittites to Egypt and then to Europe. At the time, its main use regarded nails and straps for timber roof trusses. On the other hand, Chinese were the first to use cast iron. Starting from the 6th century AD, they used it as supports for pagodas and other buildings. The subsequent important development occurred in 1856 with the filing of Bessemer patent for making steel (which is an alloy of iron and carbon). Accordingly, steel resulted stronger, and the production became faster and cheaper, boosting the subsequent Second Industrial Revolution.

3.3.1 Lead

Lead is one of the first metals known by humanity. The first evidence dates back to 7000–6500 BC, and since then, it has been continuously used due to its widespread availability and the ease to be extracted, worked with and smelted. Lead was widely used by ancient civilizations, namely, Egyptians, Assyrians and Babylonians, mostly as a sealing of iron or bronze cramps joining stone blocks. The Hanging Gardens of Babylon (today's Iraq), built by King Nebuchadnezzar II (605–562 BC) for his wife Amytis, are an important example of the use of lead in construction (Figure 3.53). Lead linings were used for waterproofing, separating the soil from the underlying stone structure and retaining the humidity level.

Records of the late Roman Empire mention the use of roof cladding made by lead, for example the old senate building in Constantinople, erected by Emperor Constantine (306–36 AD). Roofing lead sheets kept being widely used in Europe during the following centuries, but lead became also the basic component for making stained glass windows, one of the most distinctive features of Gothic Architecture (from the 12th century).

3.3.2 Iron

3.3.2.1 Wrought iron

Wrought iron is almost pure iron, having a very small carbon content. It has been used for thousands of years in applications in which a tough

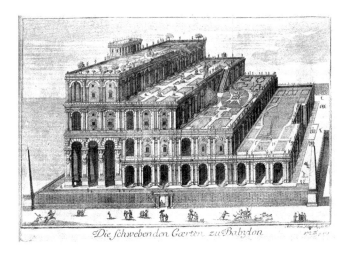

Figure 3.53 Drawing of the Hanging Gardens of Babylon (modern Iraq, 605–562 BC), one of the Seven Wonders of the Ancient World.

material was required, such as rivets, chains, railway couplings, water and steam pipes, as well as raw material for steel manufacturing. One of the most impressive examples of (surviving) wrought iron elements is represented by the Iron Pillar of Delhi (India), built by King Chandragupta II Vikramaditya between 375 and 414 AD (Figure 3.54). The pillar, 7.21 m high and weighing more than 6 tons, is made up of 98% wrought iron of pure quality, and is a testament to the high level of skills achieved by ancient Indian ironsmiths in extraction and processing.

(a)   (b)

Figure 3.54 The Iron Pillar of Delhi (India, 375–414 AD). (a) Overall view and (b) detail of the inscription on its shaft.

Regarding the manufacture of wrought iron, in ancient times, several industrial processes were available, namely, bloomery process, indirect process, and puddling and faggoting. In general, in pre-medieval times, the production of wrought iron and steel was rather expensive, mostly aimed at making tools, weapons and ornaments (exploiting the hardness and malleability of the material). Regarding constructions, iron was employed in buildings only occasionally. However, it is known that abundant wrought iron, mostly in the form of auxiliary iron tie rods, was used during the construction of Gothic cathedrals. In the case of Mallorca, all lateral naves were constructed with iron ties, that were later removed (they were cut and the end anchors left inside the masonry) after the completion of the naves (Figure 4.32). In Italy, original iron ties were used as permanent structural devices and are still visible in several Gothic cathedrals (Florence, Milan, Bologna). A remarkable example comes from India, where large wrought iron beams were used to support the ceiling of the vast stone Konark Sun Temple (also called the Black Pagoda) built in Odisha during the 13th century AD (Figure 3.55).

3.3.2.2 Cast iron

The term 'cast iron' designates an entire family of metals whose main alloying elements are iron, carbon and silicon. According to the final application, different properties can be achieved modifying the chemical composition. However, cast iron is brittle, strong under compression but not under tension, and it is not suitable for purposes where a sharp edge or flexibility is required.

With reference to Figure 3.56, starting from iron ores (rocks and minerals from which metallic iron can be economically extracted), it is possible to obtain pig iron, which is the intermediate product of smelting iron ore. In general, cast iron was made by re-melting pig iron in a cupola furnace until it liquefied, and then poured into moulds (or casts) to produce casting iron products of required dimensions.

(a)   (b)

Figure 3.55 Konark Sun Temple in Odisha (India, 13th century AD). (a) External view and (b) detail of the wrought iron beams supporting the ceiling.

(a)   (b)

Figure 3.56 Basic components for the production of cast iron. (a) Iron ore and
(b) iron pig.

Although the first furnace dates back to the 1st century BC in China, an important advancement was achieved in Europe only in the High Middle Ages (probably the late 14th century AD) when the blast furnace started to spread from Belgium. On the other hand, an example of manufacturing process during medieval times in Asia is illustrated in the Nong Shu treatise, written by Wang Zhen in 1313 AD, during the Chinese Yuan Dynasty. The illustration of Figure 3.57 depicts waterwheels powering the bellows (device constructed to furnish a strong blast of air) of a blast furnace for creating cast iron.

Up to the 18th century, coal and charcoal were abundantly used as sources of heat. As already seen for the wrought iron, also cast iron was produced thanks to blast furnaces or cupolas (Figure 3.58a). However, using coal directly led to sulphur content mingled with iron, resulting in an alloy that crumbles

Figure 3.57 Manufacturing process during medieval times according to the 14th-century
Nong Shu treatise, written by Wang Zhen in 1313 AD.

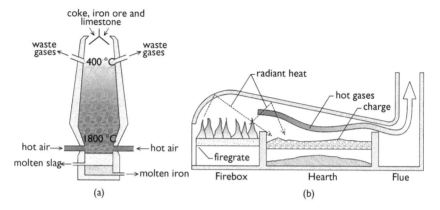

Figure 3.58 Manufacturing process after medieval times. (a) Blast furnace (cupola) and (b) reverberatory furnace.

when being worked. In order to avoid this drawback, in 1709 Abraham Darby I developed a method of producing pig iron in a blast furnace fuelled by coke rather than charcoal. Coke was obtained from the partial burn of coal, such that most of volatile materials, including sulphur, were removed. A further advantage of coke over coal was its greater porosity and greater strength.

Another attempt to enhance the cast iron manufacturing was the reverberatory furnace. According to Figure 3.58b, the material iron being heated was kept separated from contact with the fuel but not from contact with combustion gases and flames, drawn by the draught across the top of the molten iron. The term 'reverberation' is thus used in a generic sense of rebounding or reflecting the heat on the pig iron, not in the acoustic sense of echoing.

3.3.2.3 Overall comparison of iron-based products

Steel represents the technological evolution of wrought and cast iron. It is an alloy of iron and other elements, primarily carbon, widely used in construction because of its high tensile strength (and low cost). Although steel has been produced in bloomery furnaces for thousands of years, its use expanded extensively starting from the 18th century, with the invention of more efficient production methods. An important breakthrough was represented by the invention of the Bessemer process in the mid-19th century, marking the start of a new era. With the Siemens-Martin and then Gilchrist-Thomas process, mild steel replaced wrought iron.

Without describing the development of steel construction history, which belongs to the recent past, it is worth stressing the main differences between the three main products: wrought iron, cast iron and steel. Tables 3.4–3.7 give the main differences, respectively, in terms of constitutive laws, mechanical properties, advantages and disadvantages, and visual aspect.

Table 3.4 Constitutive laws for iron products

| Wrought iron | Almost pure reduction of iron ore containing only trace amounts of carbon (less than 0.1%) |

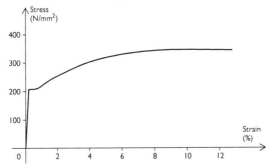

| Cast iron | Alloy of iron and carbon with carbon content below 5% |

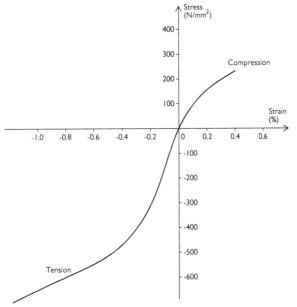

| Steel | Alloy of iron and carbon with carbon content below 2% |

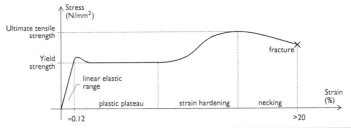

Table 3.5 Typical mechanical properties of structural grey cast iron, wrought iron and early steel (MPa)

N/mm²	Wrought iron	Grey cast iron	Steel before 1906	Steel during 1906–1968
Ultimate tensile strength (UTS)	278–593	65–280	386–494	Mild steel: 386–509
Ultimate compressive strength	247–309	587–772	=UTS	=UTS
Ultimate shear strength	≥2/3(UTS)	≥UTS	3/4(UTS)	1/√3 (practically taken as 0.6) UTS
Elastic limit (yield strength)	154–408	-	278–309	225–235
Young's Modulus (×1,000)	154–220	66–94 (Tension) 84–91 (Comp.)	200–205	205

Source: Bussell (1997).

Table 3.6 Main advantages and disadvantages of iron products

	Advantages	Disadvantages
Wrought iron	Good compressive and tensile strength 'Malleability': it can be hot-rolled or hot-hammered to produce more complex profiles and larger sections than the original flat plate Good ductility Good corrosion resistance Tough and fatigue resistant Ductility and tensile strength can be improved by reheating and reworking	The manufacturing process made small blooms (20–50 kg) that could be joined by forge welding but were never large enough to allow production of substantial structural members in one piece The working process resulted in an anisotropic 'laminar' structure with a reduced tensile strength 'across the grain'
Cast iron	High compressive strength 'Mouldability': it can be poured molten to achieve any desired shape Incombustible Good corrosion resistance Cheaper than wrought iron	Low tensile strength Brittleness Tendency to include flaws and blowholes that could reduce member strength
Steel	Good compressive and tensile strength Malleability (can be rolled into shapes) Good ductility Availability in large billets (as a result of being produced in the molten state), allowing the rolling of heavy and lengthy structural sections Ability to 'tune' its properties (in modern steels) by judicious control of carbon content and other trace elements	Somewhat inferior corrosion resistance when compared with wrought iron

Table 3.7 Visual characteristics of iron products

Visual characteristic	Wrought iron	Grey cast iron	Steel
Corners of the elements	Outer flange corners sharp, often less than 90°; 'toe' and 'root' corners rounded	External corners sharp, typically 90°; re-entrant corners rounded	As wrought iron, except for recent I-sections (sharp 90° external corners)
Cross-sectional profile	'Crisp' profile, typically •, |, I, L, T, or Z section or compound riveted section; joists and channels usually thicker than steel members	Typically 'chunky' with relatively thick sections, often ornate or complex profile (fluted or plain hollow circular or cruciform columns, |, I, ⊥ or polygonal beam sections)	Thin 'crisp' profile, typically •, |, I, U, L, or T section, solid or hollow circular or rectangular columns, or compound riveted or welded section
Flange section	Usually tapered flanges on I-sections, thickest at web; equal flange sizes Constant flange section along element	Rectangular or polygonal in beams with typically larger tension flange and small or absent compression flange Flange width or thickness may vary along element (largest at mid-span)	As wrought iron, except for recent I-sections (which have parallel flanges) Constant flange section along element
Connection methods	Rivets for all built-up sections Bolts, often square-headed Flats, bars and rods sometimes hammer-welded together in older structures Cotters and wedges for tie rods	Typically bolts (often square-headed) Beams often tied together at column heads by wrought iron 'shrink rings' fitted around cast-on beam lugs	Rivets (up to the 1950s); Bolts in clearance holes (earlier square, later hexagonal heads) Welding (20th century); close tolerance bolts (since First World War) High strength friction grip bolts (since the 1950s)

Source: Bussell (1997).

3.3.3 Metallic elements within structures

3.3.3.1 In masonry structures

Figure 3.59 shows two examples of metallic anchor ties in masonry structures, of fundamental importance to guarantee the box (or integral)

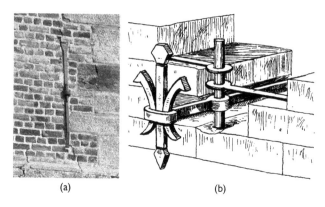

(a)   (b)

Figure 3.59 Metal anchor ties in masonry structures. (a) At the corner and (b) at the
intersection between orthogonal walls (iron strap tie).

behaviour of the buildings and increase their seismic capacity. In ancient
times, metallic cramps were used in masonry to connect blocks and provide
higher resistance against horizontal action, as shown in Figure 3.60a (see
also Sections 3.1.1 and 3.1.4). Starting with the invention of new furnaces
(from the 14th century), more complex applications were gradually developed. For instance, the façade of the Pantheon in Paris was built with an
ingenious system of iron tie bars and cramps (Figure 3.60b).

With the fabrication of beams with a given cross section (an I-beam), the
jack-arch floor became typical in many masonry buildings (Figure 3.61).
This is basically composed of a masonry arch (brick, concrete or stone)
supported on the lower flange of the beams, which are usually spaced at
a distance up to 1 or 1.5 m centre to centre. Due to the thrust of the arch,

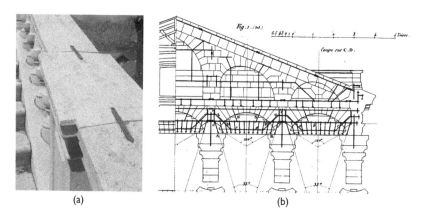

(a)   (b)

Figure 3.60 Wrought iron elements in the form of cramps and bars. (a) Example of
cramps connecting two adjoined stone blocks and (b) Rondelet's drawing of
iron tie rods and cramps in the façade of Pantheon in Paris (1756).

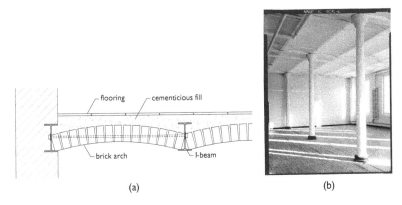

(a)   (b)

Figure 3.61 Jack-arch floor. (a) Schematic view and (b) typical aspect of the ceiling.

steel tie rods were placed to connect iron beams in wider spaced solutions. Regarding bridges, at the beginning of the 19th century, mixed cast iron-masonry structures started to be adopted. Figure 3.62 shows one of the earlier examples, the Scotch Hall Bridge over Llangollen Canal in Wales, built in 1804–1805, where the arched iron beams support the masonry structure above.

3.3.3.2 In timber structures

Metal fasteners, such as nails or cramps, have been used since Ancient Egyptians, and extensively employed by Romans. For instance, the king post trusses of Saint Catherine's Monastery built during the 6th century AD (the oldest known surviving roof trusses in the world) used an iron U-strap (or stirrup strap) connected to the post to support the horizontal tie against sagging (Figure 3.21b).

(a)   (b)

Figure 3.62 Scotch Hall Bridge over Llangollen Canal (Wales, 1804–1805). (a) View of west arch over canal and (b) detail of cast iron arches and supports.

It is since the 17th century that the strengthening of timber joints with iron started to become common. In the South of Europe, this was mainly due to wood shrinkage in low sloped king post timber roofs. Shrinkage and swelling may occur in wood when the moisture content is changed, being the former the process of reducing size, and the latter the opposite. Shrinkage occurs as moisture content decreases (e.g. elements not completely dry before being used in the construction), while swelling takes place when the moisture content increases. Usually, shrinkage induces cracking and twisting of timber elements.

Typical metallic joint plates in lattice trusses are shown in Figure 3.63, whereas modern examples of roof trusses made of wrought iron rods and timber are illustrated in Figure 3.64.

In order to strengthen common connections in timber framing, several ingenious metal sheet and metal plate devices have been manufactured along history. The most frequently used was (and still is) the joist hanger, but all of the devices shown in Figure 3.65a found extensive use, usually fixed with nails. Figure 3.65b,c exhibits two early 20th-century connectors in timber framing.

Finally, to more efficiently transfer shear at the joint, metal devices were used between timber elements (Figure 3.66). These were either provided of

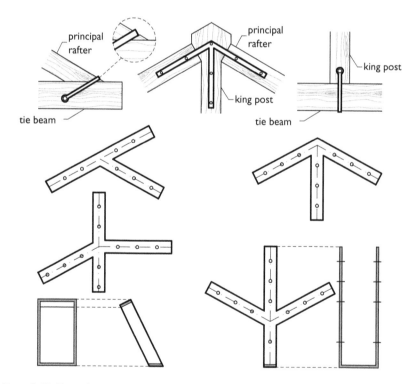

Figure 3.63 Typical metallic joint plates in timber trusses.

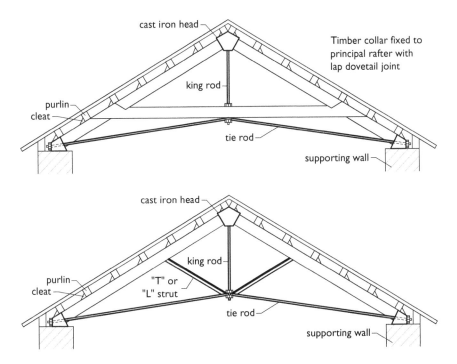

Figure 3.64 Modern examples of roof trusses made of timber and wrought iron rods, where the latter are mainly meant to substitute elements in tension.

teeth, which are then embedded into the wood of connected members or are fit into pre-cut grooves.

3.3.3.3 Lead windows and stained glass

Stained glass and lead techniques are still among the most fascinating artistic evidences in the history of construction, representing for centuries one of the distinguishing features of Gothic Architecture. Developed from jewellery, these techniques grew into a perfect mixture of art and engineering: together with the artistic skills of conceiving a suitable design, a window must support itself (gravitational load) and the action of wind (horizontal load). With this aim, robust iron frames were often required. In general, metals were used for three purposes: (1) metallic oxide powders to colour the glass (which remained further transparent), (2) soft lead strips to encircle single pieces of glass and (3) slender iron rods to reinforce the window.

Figure 3.67 shows two examples of lead windows and stained glass. In turn, Figure 3.68 displays the South Oculus at Canterbury Cathedral in England, built during the 12th century. Also, in this case, it is possible to recognize different metal elements in combination with glass: wrought iron space frame with *ferramenta* (metal framework of the window itself), grill and iron pins.

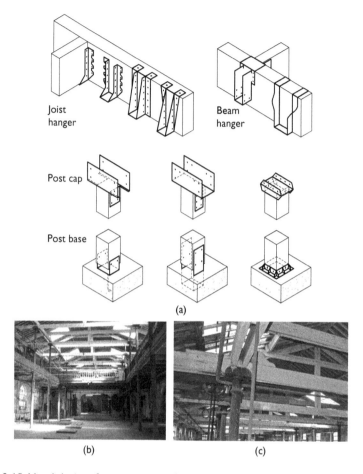

Figure 3.65 Metal devices for common timber connections. (a) Typical connections and (b and c) examples of folded and welded steel plates.

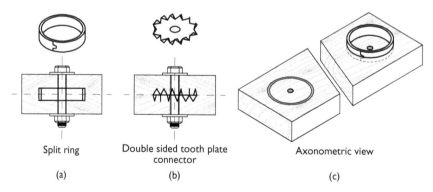

Figure 3.66 Metallic timber connectors with and without teeth. (a) Split ring, (b) double-side tooth plate connector and (c) axonometric view.

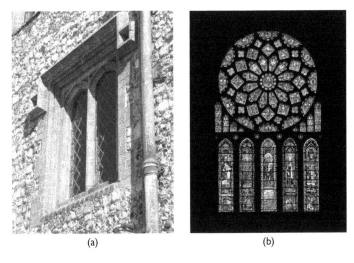

Figure 3.67 Examples of lead window and stained glass. (a) Mannington Hall, medieval country house in Norfolk (UK) and (b) original medieval stained glass in Chartres Cathedral in France (1235 AD).

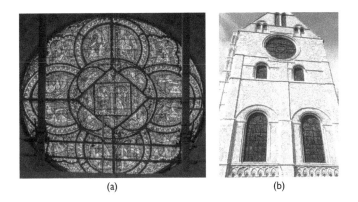

Figure 3.68 South oculus at Canterbury Cathedral in England (12th century). (a) View from the inside and (b) from the outside.

3.3.4 Metallic structures

The use of iron structural elements became a common practice only starting from the 18th century. In this regard, Figure 3.69 shows the main period of use of iron products, whereas the main constructions are described next.

3.3.4.1 Cast iron structures

It is only since the late 18th and the beginning of the 19th century that cast iron became cheap enough to be used structurally. Given the production

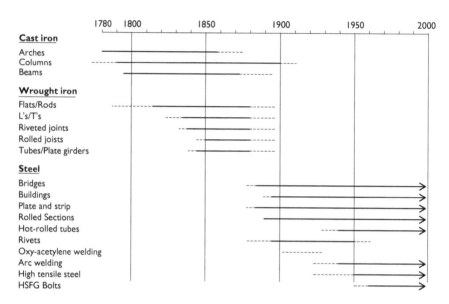

Figure 3.69 Main period of uses of iron products (Bussell, 1997).

process, cast iron elements could be of any shape, being the thickness the sole limitation. The difference in the rate of cooling through the thickness or zones, in fact, could lead to hidden defects. Due to its brittle behaviour (caused by the carbon content), cast iron was generally used for elements subjected to compression (e.g. columns and arches) rather than to bending (e.g. beams), whereas it was avoided in case of pure tensile stresses (e.g. rods). The first use of cast iron was mostly meant to substitute inflammable timber elements, frequently imitating the typical methods of jointing of timber structures. Examples of cast iron columns are shown in Figure 3.70. In particular, the decorations are worth noting, easily made by casting the metal.

Figure 3.71 presents three examples of cast iron bridges. The first one is the road bridge over the River Severn at Coalbrookdale, in England. Built in 1779, it was the first arch bridge in the world to be made of cast iron, marking the beginning of metal construction era. It was designed by Thomas Farnolls Pritchard and Abraham Darby III (grandson of Abraham Darby I, see Section 3.3.2) following the outline of a stone (and timber) vaulted bridge. It is 60 m long, with a single span of 30.5 m and a weight of around 378.5 tons.

Figure 3.71b shows the Pontcysyllte aqueduct over the valley of the River Dee, in Northeast Wales, representing the longest and highest aqueduct in Britain. Built in 1805 by Thomas Telford and William Jessop, it is 307 m long, 3.7 m wide and 1.60 m deep. It consists of a cast iron trough (38 m above the river) supported on iron arched ribs carried by 18 hollow masonry piers. Each span is 16 m long. As another example, a picture of

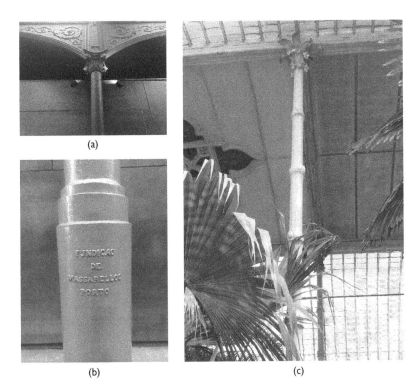

(a)

(b)   (c)

Figure 3.70 Examples of cast iron columns: detail of (a) shaft and (b) capital and decorations of the Ferreira Borges Market in Porto (1885); (c) decoration of the shaft with bamboo motif at Lednice Chateau Glasshouse in Moravia (Czech Republic, 1835).

the Liffey bridge in Dublin (Ireland) is shown in Figure 3.71c. It was built in 1816, and it is a pedestrian bridge with wooden deck, spanning 43 m with a rise of 3.35 m.

Other examples of outstanding cast iron structures are shown in Figure 3.72. The first image shows the reading room of the Bibliothèque Sainte-Geneviève in Paris (1843–1850), designed and constructed under the direction of the architect Henri Labrouste. The structure is composed by a spine of slender, cast iron columns that divide the space into twin aisles and support adorned arches. Figure 3.72b, instead, shows one of the most important constructions of the 19th century, namely, the Crystal Palace in London, built by Sir Joseph Paxton to house the Great Exhibition of 1851. At that time, it incorporated many breakthroughs, for example, it was the structure with the greatest area of glass ever seen in a building, astonishing visitors with its clear walls and ceilings, and fully embodying the spirit of British innovation that the Great Exhibition was intended to celebrate. It was 564 m long, with an interior height of 39 m, for a total covered area of 92,000 m².

Figure 3.71 Cast iron bridges at the turn of the 19th century. (a) Road bridge over the River Severn at Coalbrookdale (England, 1779), (b) Pontcysyllte aqueduct (Northeast Wales, 1805) and (c) Liffey bridge in Dublin (Ireland, 1816).

Figure 3.72 Remarkable buildings made of cast iron in France and England. (a) Reading room of the Bibliothèque Sainte-Geneviève in Paris (1843–1850) and (b) Crystal Palace in London, built to house the Great Exhibition of 1851.

3.3.4.2 Wrought iron structures

Starting from the mid-19th century, wrought iron and steel gradually substituted cast iron for three main reasons: higher performance in tension, no hidden defects and more efficient manufacturing process. The rolling mills, in particular, allowed to have elements of (theoretically) any length but with a constant cross section. Figure 3.73 shows three examples of wrought iron bridges. In particular, Figure 3.73a shows a picture of the Menai Suspension Bridge in Wales, built by Thomas Telford in 1826, which represents the world's first largest suspension bridge made of wrought iron. The length of the bridge is 304.8 m, with a central span of 176.5 m. The 16 original wrought iron chains (each weighting over 23 tons) were replaced by steel ones in 1938–1940 (Figure 3.73b).

(a)

(b)

(c)

(d)

(e)

(f)

Figure 3.73 Wrought iron bridges in France and England during the 19th century. (a) Menai Suspension Bridge in Wales (1826) and (b) detail of one of the chains; (c) Britannia Bridge in Wales (1850) and (d) original box-section element; (e) Garabit Viaduct in France (1881–1884) and (f) detail of the arch.

Figure 3.73c shows the Britannia Bridge (close to the previous one), built by Robert Stephenson in 1850 and destroyed by fire in 1970. With two main spans of 140 m each and side spans of 70 m (in total 432 m long), it was the first large wrought iron bridge of the girder type. One box-section element still lays close to the reconstructed bridge (Figure 3.73d). Finally, Figure 3.73e,f shows two pictures of the Garabit Viaduct (1881–1884) across the Truyère River in Auvergne (France). Designed and built by Gustave Eiffel as a truss two-hinged arch, the arch span is 162 m (the bridge is totally 564 m long) with a rise of 52 m.

Finally, for the World Exposition of 1889 in Paris, Gustave Eiffel designed the famous truss tower named after him (Figure 3.74). Supported by a reinforced concrete foundation, the Eiffel Tower is 300.65 m high and weighs 7,000 tons. Figure 3.74b shows the ornaments of the arch at the first level made with cast iron. The construction requested 18,000 different pieces and around 2,500,000 rivets, of which 1,050,846 were placed on-site, as

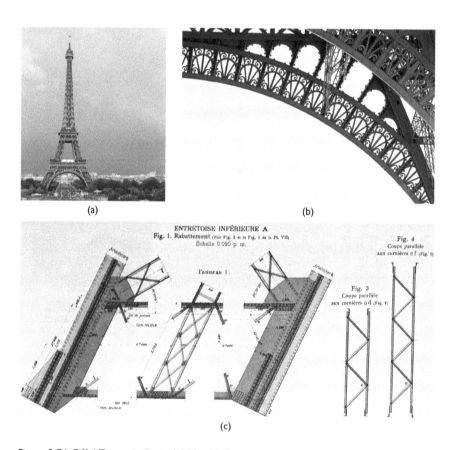

(a)   (b)   (c)

Figure 3.74 Eiffel Tower in Paris (1889). (a) Overall view, (b) cast iron ornaments in the arch at the first level and (c) detailed drawing for one of the connections.

well as detailed drawings (Figure 3.74c) for the connection and a large amount of skilled manpower.

3.3.4.3 Steel structures

As stated earlier, steel replaced wrought iron by the end of the 19th century, thanks to new methods of fabrication leading to cheaper costs for large-scale production, but also for its prime mechanical characteristics (e.g. higher ductility and capacity of undergoing cyclic loads).

Figure 3.75a shows the Eads Bridge over the Mississippi River at Saint Louis, USA (1874). Being the first use of true steel as a primary structural material in a major bridge project, the design was quite daring for that time (free span 158 m). Figure 3.75b shows the Brooklyn Bridge in New York,

(a)

(b)   (c)

Figure 3.75 Steel bridges in the late 19th century. (a) Eads Bridge over the Mississippi River at Saint Louis (1874); (b) Brooklyn Bridge, New York (1883); and (c) Forth Rail Bridge, Scotland (1890).

Figure 3.76 Victor Horta's own house in Brussels (Belgium, 1898), nowadays Horta Museum: External view and details of the Art Nouveau interiors.

USA, built in 1883. On completion, it was the largest suspension bridge in the world (its middle span is 486.3 m) and the first steel-wire suspension bridge. Figure 3.75c, finally, represents the Forth Bridge in Scotland, built in 1890. It is a railroad cantilever bridge near Edinburg, with two main spans of 521 m and a total length of 2.5 km (with approximately 58,000 tons of steel).

Figure 3.76 shows Architect's Victor Horta house in Brussels (Belgium), built in 1898, and considered the first steel building structure. It includes cast iron columns, timber elements and cut stone for the façade. Nowadays, the building hosts the Horta Museum dedicated to his life and work, considered the main pioneer of the Art Nouveau architectural style. Finally, Park Row Building in New York, USA, designed by the architect R.H. Robertson and built in 1896–1899 is shown Figure 3.77. With an overall height 119 m (30 floors above ground), it represented the highest skyscraper of the 19th century.

Figure 3.77 Park Row Building in New York, USA, built in 1896–1899.

Chapter 4

Vaulted structures in history and modern structural solutions

Compared with earlier technologies (e.g. post-and-lintel), arch and vault constructions provided the possibility of covering larger spans with limited amount of material. The deep admiration and surprise of the ancient observers for these daring elements made the way for a religious and political symbolism that have likewise developed over time.

From the constructive point of view, the most significant challenges faced by ancient masons can be synthesized as follows:

- need for extensive centrings and forms (i.e. significant consumption of material and work), or search for arch shapes and techniques requiring limited (or no) centring;
- identification of geometrical shapes that provide adequate resistance;
- need for enough buttressing to counteract the lateral thrust of arched structures.

In this chapter, the main historical developments of arch and vault constructions are revised, together with significant examples. For the sake of clarity, the main forms and the elements related to masonry vaults are presented and briefly described in the first section. The subsequent section regards the evolution of masonry vaulted structures until the beginning of the 20th century, that is, when concrete and steel became undisputed protagonists. The reader is referred to specialized publications for a more detailed description of the evolution of masonry and timber structures in history (e.g. Mark, 1990, 1993), of which some ideas are referred herein. The last section, instead, outlines the most recent techniques implemented for vaulted constructions and large roofs. For the sake of completeness, besides what is already discussed in Chapter 3, the section also includes modern solutions for simpler structural systems.

4.1 ESSENTIAL ARCH AND VAULT DESCRIPTION

Although the origin of masonry vaulted structures is lost in the mists of time, being practically impossible to date it, to many, the first example was the so-called *corbel* (or *false*) *arch*. This is an arch-like construction that uses the technique of corbeling (placing stones progressively cantilevering) to span a space or void in a structure (Figure 4.1). The corbel arch is thus constructed by offsetting from each supporting side successive courses of stone so that they project towards the symmetry axis, until the courses meet at the apex of the archway (often capped with flat stones).

Corbelling the courses means that the arches can be built with no centring or shoring. This is a peculiar feature of corbel arches, because almost all other types of arches request some kind of temporary supports. On the other hand, the horizontal courses do not allow developing any arch action (i.e. only compression stresses) until the arch is completed, requiring thus large cantilever blocks, that is, large material consumption for very limited spans. To cover larger spaces, for example ceremonial rooms, corbel vaults were often built underground so that the load on the extrados gives a stabilizing contribution, provided that the cantilever part resists to bending moment.

Moving to a traditional arch with inclined courses, it is still not clear how it developed. According to Kurrer (2008), the possible failure of upper elements of corbel arches may have led ancient builders to adopt inclined voussoirs (Figure 4.2a). Another possible development of the arch comes from a three-hinge structure, where small intermediate blocks may have been placed at the top centre (Figure 4.2b). This concept could also be at the starting point of stone lintels across the top of doors or windows.

According to the scope of the present chapter, without a discussion on the merits of each form and its origin, Figures 4.3 and 4.4 show the main types of arches. Furthermore, the main arch components are illustrated in Figure 4.5 (semicircular arch).

Masonry arches have been widely adopted also for building bridges. A schematic view is shown in Figure 4.6 together with the terminology

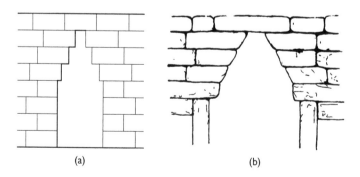

      (a)   (b)

Figure 4.1 Corbel arch. (a) Schematic view and (b) drawing of a real arch.

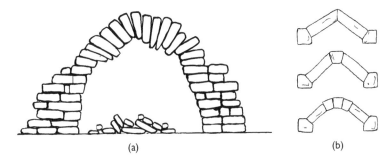

Figure 4.2 Possible origins of arches with inclined courses. (a) From the possible failure of corbel arch and (b) starting from three-hinge structures.

of selected elements. In particular, backing (or hunching) consists of a structural material (masonry or concrete), providing additional resisting volume to the main arch. Filling (or backfill) consists of non-resistant material (often non-cohesive soil or rubble) meant to provide stabilizing distributed weight and some lateral confinement. Spandrel walls are the elements between the arch and a rectangular enclosure.

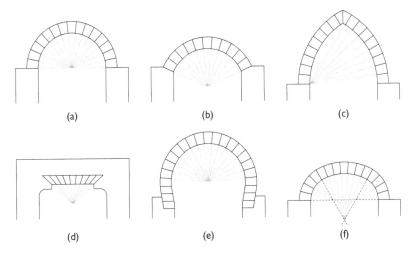

Figure 4.3 Main types of masonry arches with brief description. (a) **Roman arch** (or **semicircular** or **round arch**) is a rounded arch that forms a semicircle, (b) **Syrian** or **segmental arch** forms a partial curve, or eyebrow, over a door or window (with a slight rise), (c) **Gothic**, or **lancet**, or **pointed arch**, (d) **flat arch**, also known as **jack** or **straight arch**, extends straight across an opening (when supported, as in the drawing, on two curved cantilevers, it is called a **shouldered arch**), (e) **Moorish**, or **horseshoe arch**, extends beyond a semicircle (the top of the arch is rounded and then curves slightly before descending) and (f) **three-cantered arch** is composed of three arches with different radii and centres (these are laid out to become tangent at the connecting points).

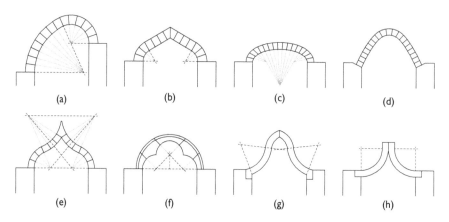

Figure 4.4 Other types of masonry arches. (a) Unequal round arch or rampant round arch, (b) tudor arch, (c) elliptical arch, (d) catenary or parabolic arch, (e) ogee arch, (f) three-foiled cusped arch, (g) reversed ogee arch and (h) inflexed arch.

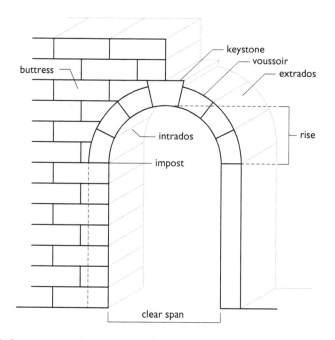

Figure 4.5 Components of a masonry arch.

Vaults share the difficulties and challenges of masonry arches during the construction process, for example need for buttressing, centring and forms. Only a few vault shapes can be constructed with very limited or even without centring, and this represented the secret of the success of Gothic and some Persian and Byzantine vaults along the history, among others.

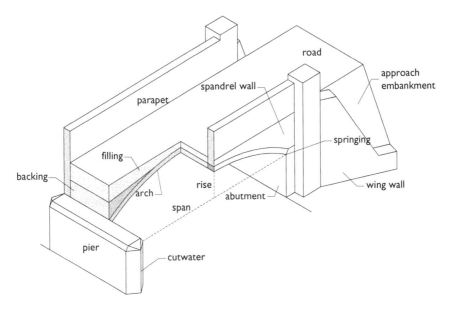

Figure 4.6 Selected masonry arch bridge terminology.

Two examples of vault building technology are shown in Figure 4.7. The first picture shows the use of rubble filling (replaceable by earth or mounds) as a temporary support for the barrel vault to be built. The second picture, instead, shows the construction process of a vault using inclined courses. The courses, leaning on the perpendicular wall, do not need any intermediate support to stand, requiring that mortar is not too plastic and relying on mortar bond, until completing the full inclined course.

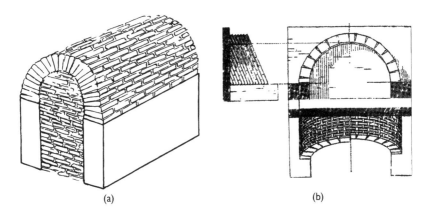

(a) (b)

Figure 4.7 Primitive vaults built without centrings. (a) Filling material as temporary support and (b) construction process with inclined courses.

Similarly, several forms of masonry vaults appeared along the history and the main types are collected in Figures 4.8 and 4.9. The latter, in particular, shows the barrel vault with lunettes (or penetrated barrel vault or underpitch) and the fan vault (in which the ribs from each springing spread out like the vanes of a fan).

Moreover, for the construction of a dome, transition elements were often necessary to pass from a square to a circular base, namely, pendentives and squinches (Figure 4.10). The former is a spherical triangle typically used to support the dome at the intersection between main nave and transept. The latter is an arch or a system of concentrically wider and gradually projecting arches, resulting in a kind of recess or niche. Placed at the corners of a square, it was also used to pass from a square to an octagonal base.

Finally, regarding cross vaults, within the countless forms developed along the history (especially starting from the 12th century AD), it is important to stress the differences between quadripartite and sexpartite cross vaults (Figure 4.11a,b). Whereas the former is obtained by dividing the bay with diagonal arches (ribs) and obtaining four webs within these ribs, the latter presents an additional division provided by a supplementary transverse arch, achieving a total of six webs (portion of the vault between the ribs).

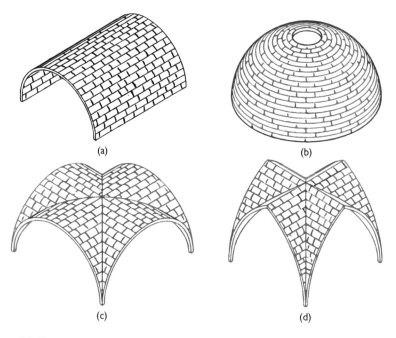

(a) (b)

(c) (d)

Figure 4.8 Examples of simple masonry vaults. (a) **Roman (or barrel) vault** (a semi-circular vault unbroken by ribs), (b) **dome** (drawn with oculus), (c) **cross vault** (Roman type) or **groin vault** (formed from intersection of two barrel vaults), (d) **Gothic cross vault** (with pointed arch).

(Continued)

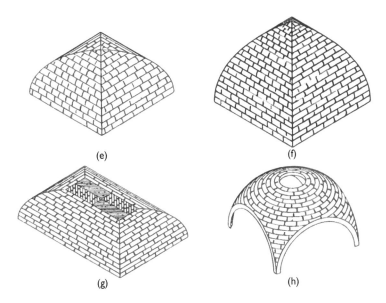

Figure 4.8 (CONTINUED) Examples of simple masonry vaults. (e) **cloister vault** (formed from intersection of two barrel vaults), (f) **cloister vault** (formed from inter-section of two pointed vaults with pointed arch), (g) **covet vault with panel** and (h) **sail vault** (drawn with oculus).

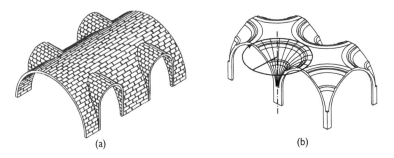

Figure 4.9 Examples of complex masonry vaults. (a) **Barrel vault with lunettes** (or penetrated barrel vault or underpitch) and (b) **fan vault**.

Figure 4.11c, in particular, shows the nave of the Cathedral of Notre-Dame de Senlis in France, where it is possible to notice the simultaneous presence of the two types of cross vaults. The advantage of the quadripartite cross vault (later development) over the sexpartite one (earlier development) may be found in the smaller thrust it generates in the longitudinal direction of the nave. This aspect is evident considering the inclination of the diagonal rib projections with respect to the longitudinal axis of the nave, that is, for a given nave width, the quadripartite vault is more compact. The smaller thrust reduces the need for temporary longitudinal buttressing during construction.

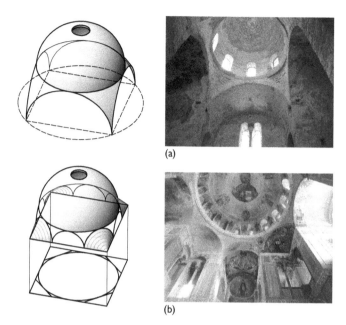

(a)

(b)

Figure 4.10 Transition elements for masonry vaults (domes in the case shown). (a) Pendentive and (b) squinch.

For the sake of clarity and completeness, the main elements of a complex quadripartite cross vault are depicted in Figure 4.12. In particular, the lateral arches are presented, where *arc doubleau* and *arc formeret* are, respectively, transversal and parallel to the longitudinal axis. Moreover, the possible ribs marking the crown are called *longitudinal* and *transverse ridge rib*, whereas *arc tierceron* is a rib extending between one corner and one ridge, and finally *lierne* is a rib not connected to any corner.

4.2 MASONRY VAULTED STRUCTURES UP TO THE 20TH CENTURY

4.2.1 Antiquity: lintels and first vaults

Historically, horizontal masonry elements have been conceived upon two alternatives:

- Lintel construction: probably inspired from timber construction and based on the combination of pillars/walls and lintels/stone slabs, the latter consisting of monolithic stones able to resist some flexural forces. Their strength stems from the material tensile strength.

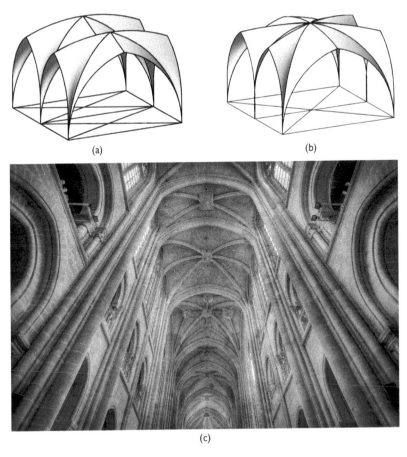

(a)   (b)

(c)

Figure 4.11 Different types of cross vaults. (a) Two quadripartite and (b) one sexpartite cross vault, and (c) simultaneous use at Cathedral of Notre-Dame de Senlis (France).

- Arched or vaulted constructions: probably inspired by natural arches and caves, their design is conceived to activate only compression forces, as their strength stems from geometry. They have represented the main roofing approach for large constructions up to the 19th–20th-century technological revolution.

Regarding lintels, the parallel between timber construction and stone lintels is evident in Greek temples (Figure 4.13), sharing the same organization and resisting principles (see also Chapter 3 and Section 4.3.3). However, despite this apparent similarity, the resistant mechanism of stone lintels is much closer to the one of jack arches. Many lintels, in fact, show cracks at mid-span, that is, a hinge at the extrados (triggered by some initial bending

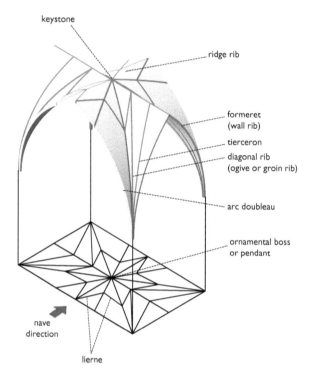

Figure 4.12 Description of a complex quadripartite vault after Ching (1995).

forces and worsened by possible environmental effects, for example a combination of thermal cycles, gravity loading, earthquakes or soil settlements), that, in many cases, do not compromise the stability of the element. This means that the lintel is working as a shallow arch leaning against the lateral piers (Figure 4.14). In this regard, Heyman (1972b) stated that lintels can work as jack arches if the piers are stable and robust enough to successfully resist their horizontal thrust. If it were not for the possibility of working as jack arches, almost all post-and-lintel ancient structures would have collapsed.

The same reasoning can be applied to other structures, such as constructions of Middle East and South America civilizations. Figure 4.15a,b shows, for instance, the colonnade of the Courtyard of Amenhotep III in Luxor Temple (14th century BC) and the drawing of one of the palaces of the city of Persepolis, capital of the Persian Achaemenid Empire (6th century BC). Figure 4.15c,d shows two pictures of the megalithic lintel of the Sun's Door in the site of Tiwanaku, Bolivia, built around 200 BC. The old drawing shows the large displacements of one of the supports and the cracks that divided the original stone in two parts that, anyhow, was standing, where the current condition is shown in the recent photo.

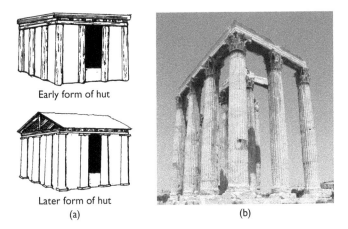

Figure 4.13 Lintel construction. (a) First Greek temples based on timber structures and (b) Temple of Olympian Zeus, Athens.

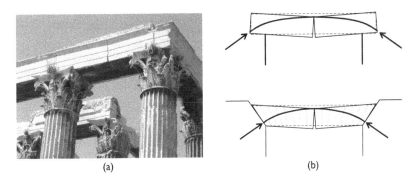

Figure 4.14 Lintel-resistant mechanism in Greek temples. (a) Crack at mid-span and (b) similarity between a lintel and a jack arch mechanism.

With the passing of time, although timber has always represented a logical, light and economically advantageous solution for floors and horizontal members, stone was still used for monumental buildings. In some cases, stone construction replicated the shapes and features of timber structures. Two examples, which date back to the 16th century, are given in Figure 4.16. Figure 4.16a is from the Red Fort of Agra (India), built around 1565–1573. In turn, Figure 4.16b shows the Royal Palace in Fatehpur Sikri (India), built ca. 1569, obviously inspired by (and even mimicking) timber construction for both arrangements and details. As it is clearly visible, the late post-and-lintel constructions show no conceptual development with respect to the first buildings, for example Greek temples. Nonetheless, the inclined elements of Figure 4.16a could be regarded as a significant improvement for reducing the span to be covered. This choice may have been the same

Figure 4.15 Ancient post-and-lintel architecture. (a) Colonnade of the Courtyard of Amenhotep III in Luxor Temple (14th century BC), (b) Persepolis, capital of the Persian Achaemenid Empire (6th century BC), (c) and (d) Sun's Door in Tiwanaku (Bolivia around 200 BC), in 1876 and 2008, respectively.

that led ancient builders (initially from Mesopotamia) to implement arch-shaped structures.

The first form of arch implemented in the history (Greek Mycenaean, Sumerian, Sasanian Persia, Pre-Columbian Mesoamerican or Khmer) was the so the so-called *false arch* implemented by means of corbelling technique (see Section 4.1). Together with the famous arch at the Lion Gate in the citadel of Mycenae (Greece, second millennium BC), one of the most impressive corbel vaults of the past is the Treasury of Atreus (or Tomb of Agamemnon) built in Mycenae around 1250 BC. In the reconstruction proposed in Figure 4.17a, it is possible to note the corbelling technique adopted for the blocks of the main dome, stabilized by the weight of the soil above it.

The same technique was adopted for the dome on squinches in the Ardashir Palace at Fîrûzâbâd in Iran (Sasanian period, early 3rd century AD), shown in Figure 4.17b. The dome lays on conical squinches supported by a square construction, with the squinches forming a zone of transition between the two geometrical entities.

(a)

(b)

Figure 4.16 Post-and-lintel constructions in the 16th century AD in India. (a) Red Fort in Agra (1565–1573) and (b) Royal Palace in Fatehpur Sikri (ca. 1569).

Other two examples of corbel vaults are given in Figure 4.18. In particular, Figure 4.18a shows one of the palaces in Palenque (Mexico) built during the 6th–7th centuries AD, together with the typical Mayan vaulting morphology (on the right). Figure 4.18b, instead, presents the vaults of Cahal Pech (Belize) and the main types of Mayan corbelled vaults (on the right). Corbel vaults were also used in the Temple of Angkor Vat (Cambodia) built during the 12th century.

4.2.2 Roman and Byzantine construction

Romans extensively used arches and vaults in their construction, achieving unparalleled levels of beauty and perfection, building impressive large-span roofs (up to 40 m) and exploring a variety of different types of vaults,

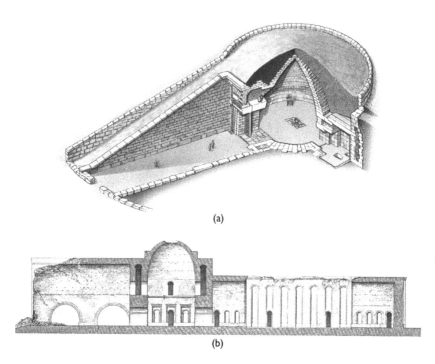

(a)

(b)

Figure 4.17 Ancient corbel domes. (a) Reconstruction of the Treasury of Atreus (Mycenae, 1250 BC) and (b) cross section of the main dome of the Ardashir Palace (Fîrûzâbâd, 300 AD).

for example barrel and cross vaults. As already discussed in Chapter 3, pozzolanic concrete played a major role in creating an almost monolithic element, being also easily adaptable to curved shapes. A key aspect, linking geometrical form and material, is that pozzolanic concrete can be cast in large masses, as it hardens by a hydraulic reaction. However, as cracks were still possible and difficult to be repaired, to avoid aesthetic inconvenience or structural problems, vaults and domes were usually supported on massive walls and foundations.

Colosseum (or *Anphitheatrum Flavium*) represents an icon itself of Roman Empire. Built in about 10 years and inaugurated in 80 AD, it could hold 87,000 people (Figure 4.19a). Erected over a massive foundation, it was made of concrete except for the external wall, which was built with dry-joint large stone block masonry (Figure 4.19b). Its shape is elliptical in plane, and it is 189 m long, 156 m wide and with an original height of almost 50 m. In spite of the massive character of the construction, along its history, the Colosseum has experienced significant alterations (after barbaric invasions and during the Middle Age) and destruction (e.g. the great earthquake of 1349) that have progressively weakened the structure. In this regard, two important stabilization interventions for the surviving exterior

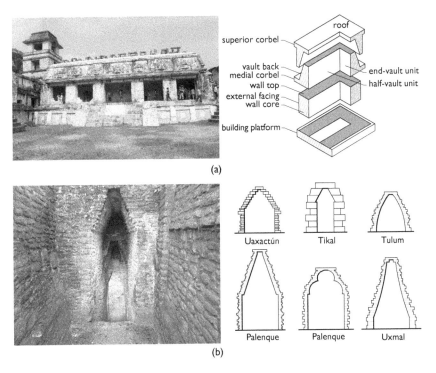

Figure 4.18 Mayan corbel vaults. (a) One of the palaces in Palenque (Mexico, 6th–7th centuries) and typical morphology and (b) Cahal Pech (Belize) and main types.

Figure 4.19 Colosseum (Rome, 72–80 AD). (a) Aerial view and (b) exterior wall.

wall were necessary during the 19th century (see Section 2.2.1). As shown in Figure 4.19b, arches play a major role in the enclosure and in the structure system itself.

Figure 4.20 shows two pictures of the ruins of Baths of Caracalla, Rome, built during 206–216 AD. The detail of one of the walls shows the concrete core (i.e. the brick or stone external leaf has been lost) with a system

(a)   (b)

Figure 4.20 Ruins of Baths of Caracalla (Rome, 206–216 AD). (a) Overall view and
(b) detail of one of the walls.

of relieving arches over the openings. Inserting brickwork arches inside
the concrete mass was a widely adopted strategy in Roman constructions
to avoid stress concentration and bending stresses (e.g. the arch upon the
opening, here further assisted by a second arch on top of the previous one).

The Basilica of Maxentius and Constantine in Rome (also known as
the Basilica Nova) represents another outstanding example of the Roman
architecture (Figure 4.21). In particular, it was the last and largest of the
great Roman Civilian Basilicas. Built during 308–312 AD, the central nave
spanned 25 m with a maximum height of 25 m. The main nave was covered by cross vaults made of concrete and supported by large buttresses
that shaped the lateral chapels. The vault thrust was channelled to the
buttressing system by means of flying arches, in perfect similitude with
the subsequent Gothic cathedrals. However, it must be stressed that the
amount of material used for this Basilica was much larger than that used

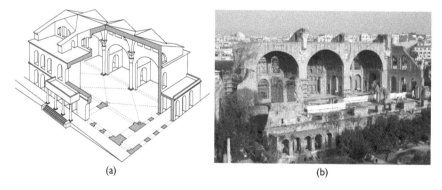

(a)   (b)

Figure 4.21 Ruins of Basilica of Maxentius (Rome, 312 AD). (a) Axonometric cross
section and (b) remains at present.

in later buildings. Figure 4.21b shows the picture of the actual state of the Basilica: the south and central naves probably collapsed during the earthquake in 847.

Pantheon is another icon of the Roman Empire (Figure 4.22a). The present shape dates back to around 125 AD, and it consists of a concrete dome of 43.3 m diameter and 43.3 m height at the oculus, over a thick concrete drum (Figure 4.22b). The thickness of the dome varies from 6.4 m at the base to 1.2 m around the oculus (with the aggregates becoming lighter when approaching to the top). For centuries, the majestic dome has been addressed as an archetype of dome constructions, representing a proven

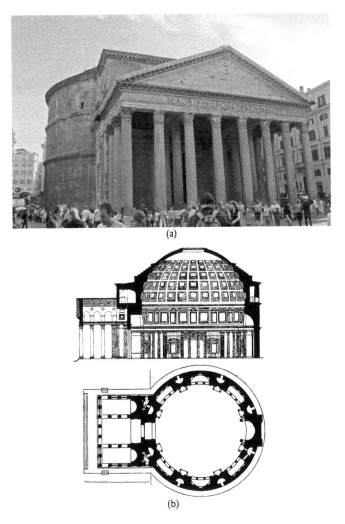

(a)

(b)

Figure 4.22 Pantheon (Rome, 125 AD). (a) External view and (b) vertical cross section and plan.

model of reachable clearance. Up to the 19th century, in fact, no masonry domes had been built with larger size.

Regarding the exterior aspect, the loss of the stone veneer permits recognizing an organized system of brickwork relieving arches embedded in the concrete mass. Figure 4.23 shows an illustration of the 16th century and a recent photo. The major damage in the building is related to the tensile stresses in the lower part of the dome. Given the low tensile strength of pozzolanic concrete, radial cracks have appeared transforming the dome in a system of (pseudo-)independent radial arches (this damage is dealt with in Chapter 6, to which the reader is referred). In this regard, it is worth to highlight the fact that the massive Pantheon drum is still able to resist the thrust caused by these discrete arches, which is larger than the constant radial pressure resulting from a true dome.

Saint Georges Church (or Rotunda of Galerius) in Thessaloniki (Greece), built in 305 AD, with an interior diameter of 24 m, 30 m height, and walls more than 6 m thick, represents a fine example of Byzantine architecture (Figure 4.24). In the shape of Pantheon, from which it borrowed the original central oculus (subsequently closed due to later uses and modifications

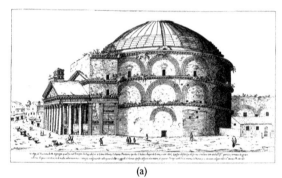

(a)

(b)

Figure 4.23 Relieving arches in the external wall of Pantheon (Rome). (a) Reproduction of the 16th century by Etienne Duperac and (b) recent photo.

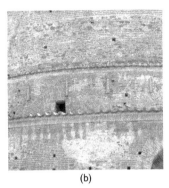

(a)   (b)

Figure 4.24 Saint Georges Church (or Rotunda of Galerius) in Thessaloniki (Greece, 305 AD). (a) External view and (b) detail of the relieving arches in the external walls.

of the original structure) and the embedded relieving arches, the Rotunda was built with bricks and local materials. Figure 4.24b shows not only the relieving arches but also the intermediate brick layers used to level and strengthen the rubble masonry.

To summarize the main features of Roman construction, a few aspects should be stressed. First, thanks to pozzolanic concrete, Roman structures were rather monolithic and massive but, at the same time, relatively sensitive and 'fragile'. Along the centuries, due to different actions (e.g. settlements and earthquakes), these characteristics contributed to cause remarkable damage or destruction. In this regard, the most successful structures (in terms of endurance) seem to have been those more highly organized and sufficiently robust, like the Pantheon. The attempt towards a more intentional structural understanding was, in fact, the Romans' strong point. This is proved by the embedded brickwork relieving arches in the wall and ribs in the vaults (a sort of hidden skeleton to control force trajectories or unload delicate parts), resulting in an overall large capacity and ductility of the building. It is also important to understand how the invention of a new material allowed both new shapes and much larger structures, linking a building style with technology.

Structural design was subsequently enhanced by Byzantines who shifted the attention to geometrical shapes and overall arrangement, rather than focusing on very large constructions. This approach led to a smart design process where the forces were adequately balanced with less material, therefore, allowing lighter structures and a lower lateral thrust. For instance, linear elements (piers or arches) that neatly follow the force trajectories were preferred to unspecialized ones such as massive walls. Similarly, ribbed domes were preferred to constant thickness ones, which, in turn, were supported on arches by means of adequate pendentives rather than continuous walls.

In general, besides the significant reduction of the amount of material, the use of this type of designed and stylized members allowed also a reduction of needed workmanship, less available than in Rome. Moreover, Byzantines did not use pozzolanic concrete, but lime concrete in rubble masonry or brick masonry. Rubble was usually combined with horizontal layers of bricks, whereas brick masonry typically shows very thick mortar joints (about 4 cm).

Figure 4.25 presents two pictures of the interior of Little Hagia Sophia Mosque (Istanbul, Turkey 527–536 AD) before the restoration, formerly the Church of the Saints Sergius and Bacchus. The pictures show the ribbed dome over an octagonal arcade.

Another great example of Byzantine architecture is Hagia Sophia (Figure 4.26a), nowadays an icon of Istanbul (Turkey). It was built between 532 and 537 AD, and the main dome has a maximum diameter of 31.2 m and a height of 55.6 m. The peculiarity of the main dome makes Hagia Sophia an impressive example of the skills of ancient builders. Originally, the dome was conceived with the same curvature as pendentives but, after its collapse due to an earthquake, it was re-built as less flat (to reduce the horizontal thrust). Except for this collapse and later local damages, Hagia Sophia has endured more than 1,000 years, successfully resisting very severe earthquakes. Sinan's additional buttressing during the 16th century (see Section 2.1) is likely to have contributed to its survival to date.

Looking from inside, Hagia Sophia is astonishing with its very vast room covered by almost suspended successive vaults. In this regard, Figure 4.26b shows a picture of the inner space. The dome is ribbed, with multiple openings at its base (where cracks are likely to develop), sustained by four large semicircular arches. The horizontal thrust of the dome and of the arches is counteracted by two half domes (and their buttressing structures) in the longitudinal direction and four large buttresses in the transversal one (Figure 4.26a,c). Accordingly, the walls that delimitate the inner space have a mere closure scope (as the dome is sustained by the circular arches).

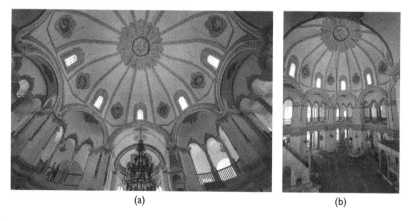

(a)   (b)

Figure 4.25 Little Hagia Sophia Mosque (Istanbul, Turkey 527–536 AD). (a) Ribbed dome supported by (b) octagonal arcade.

(a)

(b)   (c)

Figure 4.26 Hagia Sophia in Istanbul (Turkey, 532–537 AD). (a) External view; (b) view of the interior and (c) axonometric cross section, not including the large buttresses in the transversal direction.

4.2.3 Romanesque and Gothic construction

The scarceness of material and manpower in Europe during the 10th–12th centuries fostered a construction approach characterized by a certain form of 'material optimization': looking at Figure 4.27, the structures consisted mostly of usual members, such as walls and barrel vaults, in which walls had both closure and structural objectives, complemented by linear elements (e.g. buttresses or transverse arches). Three-leaf walls were normally used, being the outer leaves made of stone and the inner part of rubble masonry. Similar to Roman Architecture, the resulting massive constructions (where still a significant amount of material was needed to cover a limited span) showed, however, a certain redundancy and ductility. In particular, the structures could endure significant soil settlements and deformations without reaching collapse. In case of evident cracks, conversely to Roman concrete constructions, they could be easily repaired by filling or repointing the mortar joints.

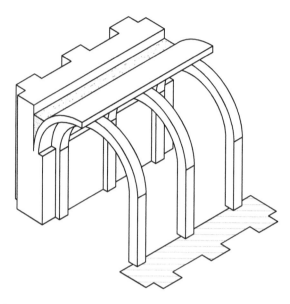

Figure 4.27 Schematic representation of the main elements adopted by Romanesque architecture.

A few centuries later, thanks to the advent of more sophisticated construction techniques, truly skeletal structures were produced, giving birth to a new building style named Gothic. Builders of that time combined and adapted elements of former architectural cultures (e.g. flying arches from Byzantine, cross vaults from Roman and other Medieval architecture, pointed arch from Middle East countries) to create structures where forces were adequately balanced and neatly transferred to buttresses and foundation. This development resulted in significant material saving (close to the minimum material consumption) as well as structural slenderness and clearance (compared with other architectural approaches). No structural 2D members exist in Gothic constructions, except for the membranes spanning across the nervures.

For the sake of clarity, the main elements of Gothic architecture are indicated on a drawing of Amiens Cathedral (France, 13th century) shown in Figure 4.28. The transverse section of the same cathedral is shown in Figure 4.29 together with the one of Beauvais Cathedral (begun in 1225). The maximum height of the nave vaults in the two cathedrals is equal to 42 and 48 m, respectively, which demonstrate how daring these structures are.

Figures 4.30 and 4.31 illustrate two impressive Gothic Cathedrals built in Spain during the 14th–15th centuries. In particular, Girona Cathedral shows the largest span (23 m) ever attained by Gothic vaults, with a maximum height of 35 m. The minutes of the two conferences of experts (held in 1386 and 1416), where the feasibility of such vaults was discussed, constitute nowadays an important document for analysing the structural choice

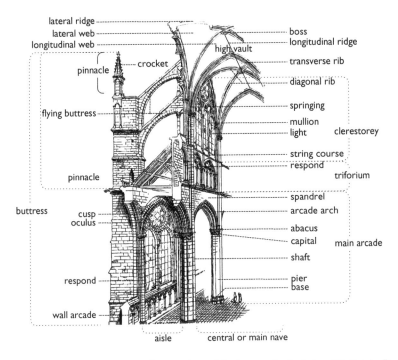

Figure 4.28 Perspective section of a nave bay of Amiens cathedral after Viollet-le-Duc.

Figure 4.29 Transverse section of Amiens cathedral (begun in 1220) and Beauvais cathedral (begun in 1225) in France.

(a)   (b)

Figure 4.30 Girona Cathedral (Spain, 14th–15th centuries). (a) View of the interior and (b) detail of the cross vaults in the main nave.

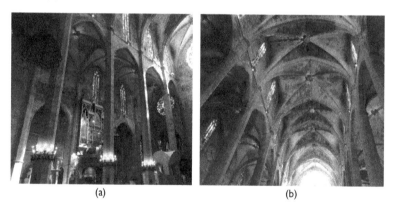

(a)   (b)

Figure 4.31 Mallorca Cathedral (Spain, 14th–15th centuries). (a) View of the interior and (b) cross vaults of the main nave.

and reasoning of ancient masons. Mallorca Cathedral, on the other hand, was considered by Mark (1982) as the epitome of Gothic construction due to its unique combination of height over ground (44 m) and clearance (the central nave vaults span 19.4 m). The extreme slenderness of the piers can only be compared with that of Jerónimos Monastery in Lisbon, as stated earlier (diameter/height ratio up to 1:15).

With respect to construction process, the fact that the structure was able to reach equilibrium (given by thrust balance) only when completed may have generated significant difficulties during the execution. In this regard, delicate stages had to be overcome, and significant temporary structures or devices may have been needed. Considering, for instance, the construction of cross vaults (i.e. the essence of Gothic), they were built using centring only for arches and ribs, whereas the shape of the webs was left to the expertise of masons (usually resulting in a double-curvature element). Figure 4.32 shows the use of iron tie rods used during the construction of Mallorca Cathedral. The tie rods were put in place to help finding equilibrium during the construction process and cut after the construction of each bay. The figure shows the remaining iron tie ends cramped in the masonry and the location of the ties in the lateral nave (second photo on top).

On the other hand, at a larger scale, new Gothic cathedrals were usually built on former Romanesque constructions, where the latter were used as provisory buttresses while gradually demolished (as for the Cathedral in Barcelona). In any case, possible damage or deformation may have been accumulated during the construction process and be still evident, as for the Mallorca Cathedral.

4.2.4 The construction of domes up to the 18th century

If the origins of cross vaults date back to Romans, masonry domes are older, having played an important role in the Middle East, especially in Persia (now Iran). Throughout Persian history, in fact, the dome was widely adopted in religious, civil and vernacular architecture, being the domed bay often used

Figure 4.32 Use of iron tie rods as auxiliary device during the construction of Mallorca Cathedral (Pelà et al., 2016).

as the constructive module. Regarding the shape, Persians preferred the pitched dome of the primitive Sasanian examples, allowing the construction without or with only limited centring. For instance, Figure 4.33 shows the Mausoleum of Sultan Muhammad Öljeitü Khudabanda in Sultaniya (Iran, 1307–1313) as it was before the restoration.

A few centuries later, impressive domes were built in Turkey by the 'Great architect' Sinan. After a careful analysis of the dome of Hagia Sophia, he replicated and even improved its original scheme. For example, he implemented techniques similar to the strengthening measures normally adopted for conservation to make new constructions, for example the perimeter iron rings at the base of the dome. This is the case of Süleymaniye Mosque built in Istanbul (Turkey) between 1550 and 1557, shown in Figure 4.34.

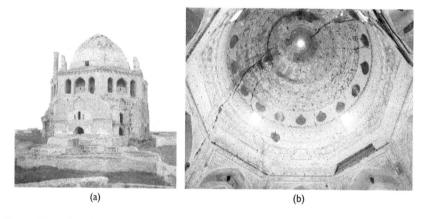

(a)   (b)

Figure 4.33 Mausoleum of Sultan Muhammad Öljeitü Khudabanda in Sultaniya (Iran, 1307–1313) before restoration. (a) External view and (b) interior view of the dome.

(a)   (b)

Figure 4.34 Süleymaniye Mosque in Istanbul (Turkey, 1550–1557). (a) Courtyard and (b) interior view of the dome.

The building is 59 m long and 58 m wide, whereas the main dome is 53 m high with a diameter of 27 m.

Sinan was also interested in the aesthetics of construction, and his search for a more homogeneous interior culminated with the Selimiye Mosque, at Edirne in Turkey (Figure 4.35). Begun in 1569, the central dome has a diameter of 43 m, and it is supported by an octagonal base. From outside, the buttressing system is barely visible.

At the same time (15th–16th centuries), Renaissance in Europe brought a renewed interest in ancient classical architecture (i.e. Greek and Roman). Accordingly, domes were used for emblematic buildings as an outstanding roofing solution, while Gothic architecture was rejected and even despised as inconsistent with classical conception. In spite of this philosophical refusal, Gothic was still prevailing due to its rationality and material optimality, integrating the lessons of classical architecture, rather than completely discarded. This ingenious synthesis can be identified in Brunelleschi's dome of Santa Maria del Fiore in Florence (built in 1419–1436, and the lantern in 1445–1461), which includes remarkable Gothic characteristics, such as the pointed geometry (to reduce the horizontal thrust) and the ribbed (and certainly complex) membrane supported on meridian arches (Figure 4.36a). These aspects were combined with classical concepts, for example the dimensions were taken from Pantheon.

Apart from the architectural styles, the dome of Santa Maria del Fiore represents nowadays one of the most complex structures ever built and still not totally understood. Brunelleschi's ingenuity allowed, without resorting to any centring or forms, building a Pantheon-like dome with a diameter of 43.6 m and a rise (interior) of 33 m, with an overall weight of 37,000 tons, over an existing masonry structure at a height of 54 m. This incredible goal was accomplished through a ribbed structure and double-layered membrane, complemented with (hardly efficient) wooden and cramped-sandstone ring ties (Figure 4.36b). The inner shell of the dome is 2 m thick and the outer is 0.6 m, whereas the ribs are more than 4 m deep.

(a)

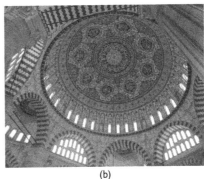
(b)

Figure 4.35 Selimiye Mosque, at Edirne (Turkey, 1569). (a) External and (b) interior view.

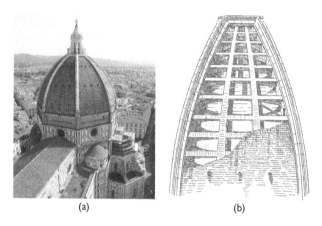

Figure 4.36 Dome of Santa Maria del Fiore in Florence (Italy, 1419–1461). (a) External view and (b) system of the dome (showing the inner structure of one of the eight slices).

Figure 4.37a shows the main geometry of slices and courses. The brick beds, in particular, are shaped as a loose string (so-called *slack line*). The reason for this is that the brick beds all lie on the surface of an inverted cone, whose axis coincides with that of the dome (the result is similar to the shape of a sharpened pencil). As a result, the concave geometry of the

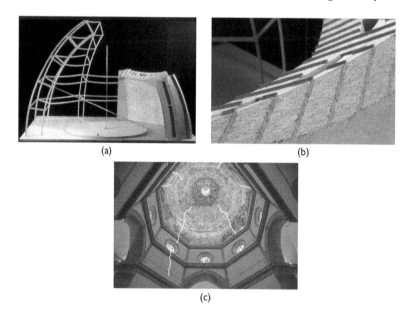

Figure 4.37 Dome of Santa Maria del Fiore in Florence (Italy, 1419–1461). (a) Geometry of the slices and brick beds, (b) detail of the 'spinapesce' and (c) present crack pattern.

membranes permits a better interlocking at the connections when compared with straight courses. Moreover, the use of the herringbone bond of bricks (*a spina di pesce*) was a key element to provide grip between brick layers during construction (Figure 4.37b). For details on the construction process of the dome, the reader is referred to videos from National Geographic (2014) and Nova PBS (2015). Again, given the daring approach and large size, the structure has been monitored for more than 60 years (Ottoni and Blasi, 2015), and the actual cracked condition is depicted in Figure 4.37c.

Brunelleschi's solution was inspirational for another masterpiece of the Italian Renaissance, namely, the dome of Saint Peter's Basilica in Rome (Italy, 1564–1590). The design process underwent several changes under the supervision of eminent architects of that time, as Michelangelo and Domenico Fontana. The final design was provided by Giacomo della Porta with clear references to Brunelleschi's innovations, for example double-layered membrane, ribbed structure and parabolic profile with a lantern. Again, taking the Pantheon as model and dimensional reference, the interior diameter of the dome is 42.3 m, even if it rises 120 m above the ground (Figure 4.38).

Soon after its construction, starting from the 1630s, cracks began to develop and gradually grew over time to the extent that, by the mid-18th century, damage was rather evident. Long meridian cracks were visible running along the dome intrados (Figure 4.39a), and the alarm spread across the entire Europe. Remedial action needed to be taken. In 1742, Pope Benedict XIV appointed a committee of scientists, known as 'The Three Mathematicians', to report on the condition of the dome. The scientists, through two subsequent reports, stated the dome was seriously damaged and it would require extensive strengthening operations.

Since other intellectuals gave controversial opinions, Pope Benedict XIV decided to seek the advice of another brilliant scholar. The Chair in Experimental Philosophy at the University of Padua and director of the new materials testing laboratory, Giovanni Poleni, was asked in 1743 to

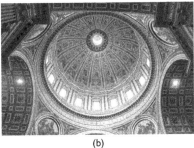

(a)   (b)

Figure 4.38 Dome of Saint Peter's Basilica in Rome (Italy, 1564–1590). (a) External and (b) interior views.

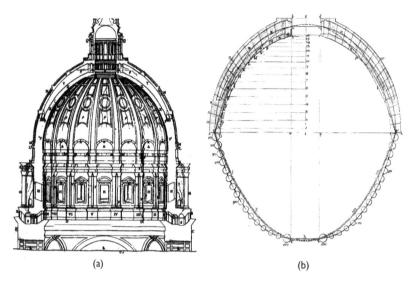

(a)   (b)

Figure 4.39 Dome of Saint Peter's Basilica in Rome (Italy, 1564–1590). (a) Crack pattern and (b) Poleni's experiment for ascertaining the stability of the dome.

assess the condition of the dome. In a preliminary report of the same year, applying the catenary principle, Poleni proved Saint Peter's dome to be adequately designed, stating that the radial cracks (existing in almost every dome) were by no means connected to problems associated to collapse. In his final report (Poleni, 1748), he explained his analyses in an original and meticulous way, also by means of experimental activities (Figure 4.39b). With the cooperation of Vanvitelli between 1743 and 1748, Poleni proposed the strengthening of the dome by placing a set of iron rings located at the dome springing.

In subsequent centuries, the evolution of domes saw a progressive lightening of the structure (i.e. less amount of material, thus less weight and less thrust) while preserving and even enhancing its symbolism: the interior central and clear space crowned by a celestial roof, and the external outstanding and visible landmark. This goal was achieved by designing systems composed of multiple superposed domes (normally two or three), of which only one was usually fully structural. Starting from the 17th century, similar solutions were implemented in Persia, India (influenced by Persian architecture) and Europe (possibly influenced also by the Persian Architecture).

Figure 4.40a shows an example of a cross section of a Persian dome of that time. It is possible to observe the usual and appealing external shape, this time composed of a sophisticated structure with masonry stiffeners and timber, providing a higher external volume and a pitched inner dome. Many other hidden systems exist, including masonry stiffeners and

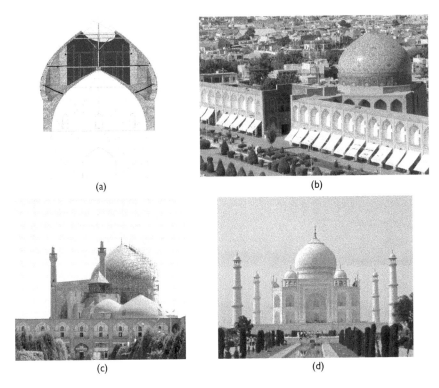

Figure 4.40 Persian domes in the 17th century. (a) Example of a cross section with
the hidden timber and masonry stiffeners' structure, (b) Sheikh Lotfollah
mosque in Isfahan (Iran, 1603–1619), (c) Masjid-e Shah (Royal Mosque) in
Isfahan (Iran, 1611–1638) and (d) Taj Mahal in Agra (India, 1631–1653).

columns, or timber structures or a combination of both, to stabilize two
domes, being not always clear which are the structural elements and which
are the elements used as auxiliary devices in the construction process.
Figure 4.40b,c shows the pictures of two mosques in Isfahan (Iran) and
their domes: the Sheikh Lotfollah mosque built during 1603–1619, and the
Masjid-e Shah (Royal) mosque built between 1611 and 1638 with an inner
diameter of 12 m. Figure 4.40d shows the Taj Mahal, in Agra (India), built
during 1631–1653. A hidden dome with a diameter of 17.70 m and a total
rise of 24.4 is surmounted by an outer shell 61 m in height. In this case, the
Persian influence is clear.

In the same period, in Europe, probably for the first time in history,
Christopher Wren, in cooperation with Robert Hooke attempted to design
a structure according to a scientific reasoning (i.e. the catenary principle)
rather than traditional rules (of thumb). They worked on Saint Paul's
Cathedral in London (Figure 4.41a), built during 1677–1708, where the
inner dome is 65 m above the floor. Hooke's structural insights might have

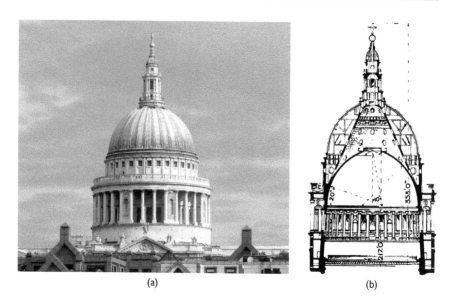

(a)   (b)

Figure 4.41 Saint Paul's Cathedral in London (1677–1708). (a) External view and
(b) cross section with the hidden structural conical dome, together with
the visible internal and external semicircular domes.

led to a new and more economical way of producing an emblematic dome.
As shown in Figure 4.41b, the true structure consists of an almost conical
brick masonry dome (probably designed by Hooke, with the idea of approx-
imating a catenary), whose shape is able to support the outer shell and the
lantern on the top (whose weight is 8,500 tons). Lighter domes, with mostly
decorative roles, are below and above the cone. Similar to Persian domes
(see Figure 4.40a), the external dome is supported by a timber structure
resting on the conical dome.

This multiple dome arrangement became a standard solution during the
18th century. Two additional examples are shown in Figure 4.42. Very
similar to Saint Paul's Cathedral, the first picture shows the Panthéon
in Paris (France, 1758–1789), where, to reduce the structural weight,
ancient builders resorted to masonry reinforced with iron. As explained in
Section 2.3, iron corrosion tends to cause severe damage in the long term,
which was the case for this dome. The second picture is from the United
States Capitol dome, built in Washington in 1793 with the same double
dome system. The large outer dome, in particular, is a thin shell held up
by a ring of curved iron ribs. The smaller dome underneath, instead, is
self-supporting and visible only from the inside. The U.S. Capitol dome
deserves a place in the history of construction also because it was one of the
earliest domes made of prefabricated cast-iron ribs. This innovation greatly
reduced the weight of the structure, allowing easy and fast construction of
domes around the world.

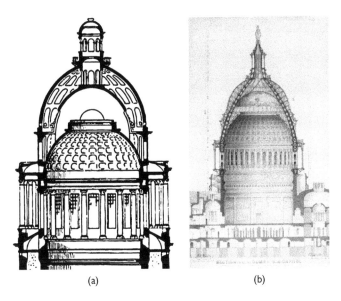

(a)   (b)

Figure 4.42 Two domes following the scheme of Saint Paul's Cathedral in London. (a) The Panthéon in Paris (France, 1758–1789) and (b) United States Capitol (Washington, 1855–1866).

4.2.5 The epitome of masonry large-scale vaults

The epitome of masonry large-scale vaults, and analogously, of large-scale constructions made of unreinforced masonry, is linked to the unique personalities of Alessandro Antonelli (1798–1888) and Antoni Gaudí (1852–1926). More recently, also the works of Rafael Guastavino (1842–1908) and Eladio Dieste (1917–2000) are receiving much attention, even if not addressed here. Their projects reveal a deep knowledge of construction history and techniques as well as a sound understanding of brick and stone masonry structural possibilities. With no doubts, their complex and daring structures are the result of a genuine investigation.

In this regard, little is known about Antonelli's design approaches. According to the extraordinary results achieved that challenge the static strength of masonry, he might have adopted some scientific method (perhaps 3D hanging models). Regarded as the most representative Antonelli's works, Figure 4.43 shows the pictures of the dome of the Mole Antonelliana in Turin (Italy), 167 m high, built with brick masonry (1863–1889), and the dome of Basilica of San Gaudenzio in Novara (Italy), built in 1840–1873 and 125 m high.

The catenary theory was at the base of Gaudí's projects. Being probably the first architect to do so, he widely referred to 3D hanging models (or scaled plaster models) as a way to apply the catenary principle to true 3D structures. For instance, the historical photo of the hanging model used

(a)   (b)

Figure 4.43 Main Antonelli's works. (a) Mole Antonelliana in Turin (Italy, 1863–1889) and (b) Basilica of San Gaudenzio in Novara (Italy, 1840–1873).

for the Church of Colònia Güell is presented in Figure 4.44a. Designed in 1908, the works started in the same year, but they were interrupted in 1914 when only the crypt (ground level) was completed. Figure 4.44b, instead, shows the reconstruction of the plaster large-scale model of the nave of the Temple of Sagrada Família in Barcelona (the original got damaged during the Spanish Civil War). It is worth noting, again, that, to validate the equilibrium and the stability of stone structures, scale does not represent an issue

(a)   (b)

Figure 4.44 Gaudí's models. (a) Hanging surfaces for the Church of Colònia Güell in Barcelona (Spain) and (b) modern reconstruction of the scaled model for the Temple of Sagrada Família in Barcelona (Spain).

as the catenary principle is not affected by size effect (i.e. stresses within the structure are not relevant). This is not the case for fracture processes or dynamic loading, in which the results are size dependent. Nowadays, the church is still under construction (expected to be completed in 2026) and, conversely to Gaudí's design, reinforced concrete is being widely used.

Only based on the catenary principle and funicular surfaces, Antonelli and Gaudí showed that masonry could stand as a material for modern architecture, with advantages regarding constructability, economy and aesthetics. Their structures are light, vast in dimensions and slender, with a limited material consumption (in relation to their daring dimensions), with skeletal arch systems determined by the load paths and balance of thrusts. Their works are also characterized by a strong aesthetic value, reaching in the case of Gaudí the highest creativity, as an amazing example of synthesis between art and technique, shape and strength.

The unceasing innovative masonry structural development that started with some Byzantine and Ottoman constructions, subsequently boosted by Gothic architecture, came to an end when masonry was replaced by steel and reinforced concrete as the main construction materials. Consequently, framed systems became the typical construction technique during the 20th and 21st centuries. Several engineers and architects have revisited the possibilities of building shells and resistant-by-shape structures in concrete (Nervi, Candela, etc.) and unreinforced (Guastavino) or reinforced masonry (Dieste), as stated earlier. In doing so, they have produced much interesting and valuable structures of the 20th century.

4.3 MODERN STRUCTURAL SOLUTIONS

Starting in the late 19th century, new materials, structural types and building technology began to spread, being still widely adopted. In particular, besides metallic structures and steel, the most important are:

- reinforced concrete: developed across the 19th and during the 20th century, it became the most used structural material in construction;
- prestressed concrete: developed by Eugene Freyssinet in the 1930s, it extended the range of application of concrete structures to large spans, competing with steel structures (see Section 4.3.3 for further details);
- glued and laminated wood (glulam): it allowed timber structures to surpass the limitations imposed by the size (length and cross section) of the trees, more recently extended from linear members to cross laminated timber panels, for walls, floors and a range of new possibilities (see Section 4.3.3 for further details);
- textiles and synthetic materials: used for membranes, tensegrity and pneumatic structures.

Accordingly, the following structural types can be identified:

- trusses;
- post and lintel;
- frames;
- arches and classical vaults;
- shell structures;
- suspended structures.

In the following, arch bridges are addressed first, followed by vaults and domes (i.e. large-span roofing system), focusing on the main types used in the last decades for both typologies. Then, the other structural systems are briefly revised.

4.3.1 Arch bridges

Taking advantage of the vaulted shape, since the mid-18th century, many large-span bridges have been built in concrete and steel. Without claiming to describe the history of arch bridges, three main typologies may be distinguished, and a few examples are given ahead:

- **Deck arch bridge:** this type of bridge comprises an arch where the deck is completely above the arch. Converse to masonry bridges where the spandrel is solid (i.e. closed-spandrel deck arch bridge), for concrete and steel bridges, the deck is supported by a number of vertical columns rising from the arch (i.e. open-spandrel deck arch bridge).
- **Tied-arch bridge (or bowstring arch):** this type of arch bridge incorporates a tie between two opposite ends of the arch. The tie is usually the deck which withstands the horizontal thrust forces of the arch (normally at the abutments). The deck is normally suspended from the arch.
- **Through arch bridge:** this type of bridge has an arch whose base is below the deck, but whose top rises above it, so the deck passes through the arch, which can also work as a tie. The central part of the deck is supported by the arch via suspension cables or tie bars, whereas the ends of the bridge may be supported from below, as with a deck arch bridge. Any part supported from arch below may have spandrels that are closed or open.

Deck arch bridges are presented first. Figure 4.45 shows two impressive arches designed by Italian Engineers in the 1960s. The first picture regards the deck arch bridge named Bisantis Bridge in Catanzaro (Italy) designed by Riccardo Morandi in 1958 and inaugurated in 1962. The arch has a

(a)

(b)

Figure 4.45 Italian reinforced concrete bridges. (a) Bisantis Bridge in Catanzaro (Italy, 1959–1962) and (b) Musmeci Bridge in Potenza (Italy, 1971–1976).

span of 231 m and a rise of 66 m. At that time, among the concrete bridges, it was the highest in the world, being only the second one in terms of span, with a costly formwork of about 120 m high from the lowest part of the valley. Figure 4.45b, instead, shows the Musmeci Bridge in Potenza (Italy), designed by the Sergio Musmeci in 1967 and built between 1971 and 1976. Following the seminal idea of a deck arch bridge, it is made of only one membrane of reinforced concrete (about 30 cm thick) moulded to form four contiguous arches (with a total length of 560 m). The concrete sheet is shaped into a 'finger-like' structure (according to the designer, 'the best form to convey the forces'), which supports the whole bridge, and it is also used as a pedestrian walkway.

As far as deck arch bridges are concerned, another distinction can be drawn: stiff arch/slender deck and laminar (slender) arch/stiff deck. Regarding the former, and the most used system, Figure 4.46a shows the Krk bridges, built in 1980 between the islands of Sveti Marko and Krk in Croatia. The two spans are 390 and 244 m, whereas the rise of the arches is 67 m. On the other hand, Figure 4.46b presents an example of the opposite scheme. This is the Infante Don Henrique Bridge, built in 2002 in Porto (Portugal), over the Douro river. Arch span is of 280 m, whereas the rise is of 25 m.

(a)

(b)

Figure 4.46 Reinforced concrete arches. (a) Krk bridges between the islands of Sveti Marko and Krk (Croatia, 1980) and (b) Infante Don Henrique Bridge in Porto (Portugal, 2002).

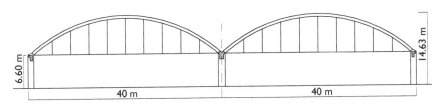

Figure 4.47 Precast tied arches for the roof of a bus garage in Barcelona (Spain, 1965).

Figure 4.48 The CFST Wushan Yangtze River Bridge (China, 2005).

An example of a tied arch is shown in Figure 4.47. The picture shows the precast tied arches (spanning 40 m each) for the roof of a bus garage in Barcelona (Spain, 1965). The arch cross section is only $0.20 \times 0.45 \, \text{m}^2$, whereas the ties were built in precast pre-tensioned concrete. In pre-tensioning, the reinforcement, in the form of tendons or cables, is stretched before the concrete is placed (see Section 4.3.3).

Finally, a typical example of through arch bridge is given in Figure 4.48 showing the Wushan Yangtze River Bridge, one of the largest bridges in the world. This was constructed with concrete-filled steel tubes (CFST), and the main arch is composed of a 3D truss, allowing a span of 460 m. CFST has been widely implemented in large-span bridges in China since 1990, and more than 33 CFST arch bridges with a span larger than 200 m have been built in China until 2007.

Regarding short spans, besides steel and concrete, glulam is also used for building slender arches for bridges. Figure 4.48 shows the roof of the Richmond Olympic Oval (Canada) with a span of 100 m; the three-hinged Keystone Wye arch bridge in South Dakota (USA, 1968), spanning 47 m, is shown in Figure 4.49.

(a)   (b)

Figure 4.49 Large-span arches made with glulam. (a) Internal view of the Richmond Olympic Oval in Canada (note that steel arches were also used) and (b) the three-hinged Keystone Wye arch bridge in South Dakota (USA, 1968).

4.3.2 Vaults and domes

4.3.2.1 Curved network systems and folded shells

Figure 4.50 illustrates the main schemes for the construction of shells by means of curved network systems. An interesting example is represented by the Tacoma Dome in Washington (Figure 4.51), built in 1983 with glulam elements. It has a diameter of 161.5 m, and it is the world's second largest arena with a wooden dome. The largest one is the Superior Dome in Marquette (Michigan, USA) with a diameter of 163.4 m.

Regarding folded shells, Figure 4.52 shows the main schemes of quasiflat and vaulted roofing systems. In this regard, the Hangar at Orly (France, 1921–1940), designed by Eugène Freyssinet (1879–1962), inventor of the prestressing in concrete structures, is shown in Figure 4.53. The cross section of the arches is a thin shell structure built with a movable scaffolding (Figure 4.53b).

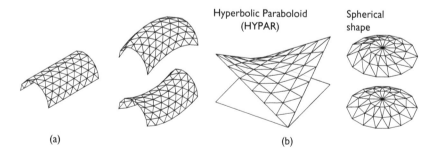

(a)   (b)

Figure 4.50 Vaults and domes: main schemes of curved network systems. (a) Simple curved system and (b) double curved systems.

(a)   (b)

Figure 4.51 Tacoma Dome in Washington (USA, 1983). (a) During the construction stages and (b) current external view.

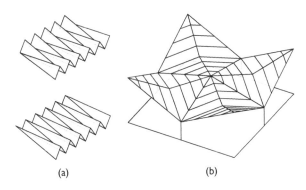

(a)   (b)

Figure 4.52 Vaults and domes with folded shells: schematic views of (a) Quasi-flat and (b) vaulted roofing systems.

4.3.2.2 Double-curvature shells

Figure 4.54 illustrates the main schemes for double-curvature shells, used by engineers and architects starting on the 20th century to create impressive structures. Among others, Nicolas Esquillan, Félix Candela, Heinz Isler, Eeron Saarinen, Pier Luigi Nervi, Eduardo Torroja, Jorn Utzon and Oscar Niemeyer can be recalled.

The *Centre des Nouvelles Industries et Technologies* (CNIT) in Paris is the world's largest reinforced concrete shell in terms of square footage of area covered per support (Figure 4.55). Built between 1956 and 1958 on the basis of Esquillan's project, it has a total covered area of 22,500 m², a free height of 46.3 m and a distance between the supports of 218 m.

Three works of Félix Candela (1910–1997) are shown in Figure 4.56. In the *Los Manantiales* Restaurant in Mexico (1958), Candela wanted a building integrated in the gardens, as a lotus flower floating on the water. The result was the design of a lily-like circular dome, formed by the intersection

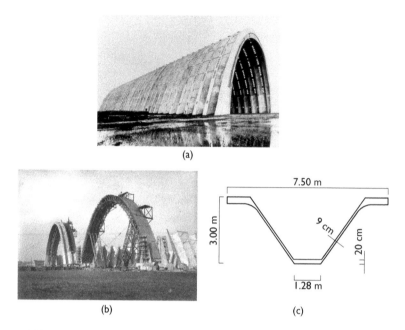

Figure 4.53 Hangar at Orly (France, 1921–1940). (a) External view, (b) construction phase and (c) thin shell cross section.

Figure 4.54 Vaults and domes: main schemes of double-curvature shells.

Figure 4.55 CNIT in Paris by architects Robert Camelot, Jean de Mailly, Bernard Zehrfuss and Jean Prouvé (France, 1956–1958). (a) External and (b) internal views.

Figure 4.56 Arch. Félix Candela's main works. (a) *Los Manantiales* restaurant (Mexico, 1958), (b) structure in the Aquarium of Valencia (Spain, 2001) and (c) *Chapel of Nuestra Señora de la Soledad* (Mexico, 1955).

of eight segments from four hyperbolic paraboloids (surface similar to that of a saddle horse). The external diameter is 42 m, with a maximum height of 8.25 m that is reduced inside to 5.90 m. A very similar structure, almost a replica, was built with steel fibre reinforced concrete thin shells for the Aquarium of Valencia in 2001 (Figure 4.56b). The last picture is the Chapel of *Nuestra Señora de la Soledad* in Mexico (1955), with a rhomboid plant covered with a concrete straight-edged hyperbolic paraboloid, where the tips to the longer sides are raised.

Heinz Isler's (1926–2009) main works are shown in Figure 4.57. In particular, the petrol station at the Bern–Zurich motorway and the indoor swimming pool in Brugg, both in Switzerland. The Kresge Auditorium at the Massachusetts Institute of Technology designed by Eeron Saarinen (1910–1961) is shown in Figure 4.58.

(a)   (b)

Figure 4.57 Eng. Heinz Isler's main works. (a) Petrol station at the Bern-Zurich motor-way and (b) indoor swimming pool in Brugg, Switzerland.

Figure 4.58 Arch. Eeron Saarinen's Kresge Chapel in Massachusetts (USA, 1953).

An important role in the modern history of constructions was played by Eduardo Torroja (1889–1961) and Pier Luigi Nervi (1891–1979). The deep knowledge of concrete, supported by experimental activities on scaled or full-scale prototypes, allowed the two engineers to build masterpieces. One of Torroja's major creations was the roof of the *Frontón Recoletos* (Spain, 1935), unfortunately destroyed during the Spanish Civil War in 1939 (Figure 4.59). The roof was a unique two-lobe thin shell (radii 12.2 and 6.4 m) that covered a surface of $55 \times 32.5\,m^2$ and, in those areas where skylights were needed, the shell was replaced by a triangulated structure designed for the insertion of glass panes. The thickness of the shell was only 8 cm, except at the connection between the cylindrical sectors, where it was increased to 30 cm, to resist the transverse bending moments and to adequately cover the reinforcement bars found there.

Another impressive roof was the one for the grandstand of Zarzuela's Hippodrome begun in 1935 (Figure 4.60). Conceptually, to combine the most appropriate structural solution with architectural aesthetics (in this

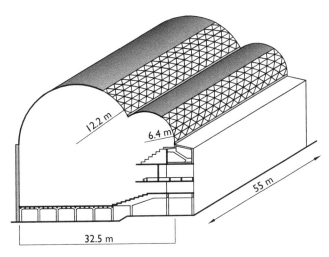

Figure 4.59 Eng. Eduardo Torroja's Frontón Recoletos (Spain, 1935): axonometric view.

Figure 4.60 Eng. Eduardo Torroja's Zarzuela's Hippodrome (Spain, 1935).

case strictly related to the rationality of the design), Torroja moulded the form of a simple flat roof until achieving a hyperboloid of variable thickness. The reinforced concrete shell thickness, thus, varies from 65 cm in the zone of the pillars to 5 cm on the ends of the cantilevers, without ribs but with a rear tie beam. The cantilever measures 12.8 m with an effective depth of 1.50 m.

Similar to Torroja, Nervi proposed important projects inspired by static and constructive forms. Among others, the Italian Air Force Hangar in Orbetello (Italy) built between 1939 and 1941 and shown in Figure 4.61a, later blasted by the German army during their retreat in early 1944. The hangar roof had a parabolic shape to better accommodate the gravitational loads, and this shape led to a low amount of steel reinforcement. In order to lighten the structure, the ribs were made of truss elements precast on the ground and, hoisted into place by a crane, joined together by a poured-in-place connection (a few arches were solidly cast to provide additional stability to the system, along with a continuous edge truss). The lighter roof also enabled a simple support system made of six isolated large buttresses. The roof covered an area of $100 \times 36\,m^2$ with a total height of about 25 m.

If precast concrete represented an innovative technique for the Italian Air Force Hangar, the Turin Exhibition Hall (Italy, 1947), shown in Figure 4.61b, is an impressive example of the so-called *ferro-cement technique*. Originated in the 1840s in France, this is a composite material composed by high-quality mortar or plaster (lime or cement, sand and water) applied over numerous layers of steel mesh (diameter in the range of 0.5–1.5 mm) and a few bars. Accordingly, it is possible to construct relatively thin, ductile, crack resistant and strong surfaces of any shape. With an almost square plan, two side galleries and a semicircular apse, the hall covers approximately $200\,m^2$. In particular, the large $96 \times 75\,m^2$ hall has a thin, corrugated vaulted roof made of a series of ferro-cement components structurally bound to each other. The installation took place using a special scaffolding with reinforced concrete ribs along the ridges and valleys of the roof section. The wave-shaped elements are 4.40 m in length with a thickness of 4 cm.

Another example of Nervi's creativity is the *Palazzetto dello Sport*, which is an indoor arena built for the Summer Olympic Games hosted in Rome in 1960, see Figure 4.62. The arena is built with a ribbed concrete shell dome with 61 m in diameter braced by concrete flying buttresses. Since much of the structure was prefabricated, the dome was erected in just 40 days.

From Figures 4.63 to 4.66, other well-known examples of curved roof structures are presented.

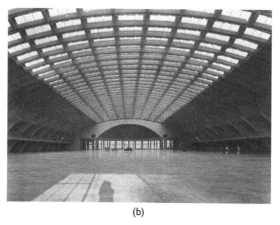

(a)   (b)

Figure 4.61 Eng. Pier Luigi Nervi's main works. (a) Italian Air Force Hangar in Orbetello during the construction phase (Italy, 1939–1941) and (b) Turin Exhibition Hall (Italy, 1947).

(a)

(b)

Figure 4.62 Eng. Pier Luigi Nervi's *Palazzetto dello Sport* (Rome, 1958–60). (a) External photo of the construction yard and (b) internal view.

4.3.2.3 Suspended, pneumatic and membrane structures

Figure 4.67 illustrates the main schemes for domes and vaults made by suspended structures. The drawings on the left represent typical cases where double cables are implemented, either in a concave or convex configuration. Being the cables prestressed, in the first case, the internal vertical elements are compressed (struts), whereas, in the second case, they are stretched (ties). In turn, the schemes on the right show roofing systems based on a single cable with an overall double curvature of the structure. Since the

Figure 4.63 Opera House in Sidney (Australia, 1957–1973) by Arch. Jorn Utzon (who resigned in 1966) and structural engineering by Ove Nyquist Arup, built with prestressed concrete ribbed shell.

(a)   (b)

Figure 4.64 Palau Blaugrana in Barcelona (Spain, 1971) by Arch. Francesc Cavallé and Josep Soteras. (a) External and (b) internal views of the ribbed dome.

Figure 4.65 Church of Saint Francis of Assisi in Pampulha (Brasil, 1942) by Arch. Oscar Niemeyer: RC shells exhibiting short durability, as bars have a thin cover and concrete compaction was rather difficult.

(a)   (b)

Figure 4.66 Saint Louis Lambert International Airport TI (1956). (a) Western perspective and (b) internal view.

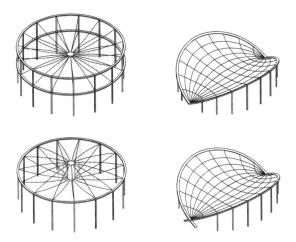

Figure 4.67 Vaults and domes: schematic views of suspended structures.

cables only work in tension, the external ring (to which they are connected) is basically compressed. In the following figures, some examples of real case studies are shown.

Figure 4.68 shows the external and internal views and of the Dorton Arena, the first structure in the world to use a cable-supported roof. Its design features a steel cable supported saddle-shaped roof in tension, held up by parabolic concrete arches in compression. The arches cross about 6 m above ground level and continue underground, where the ends of the arches are held together by more steel cables in tension.

Images of the Madison Square Garden Arena are shown in Figure 4.69. Also in this case, the roof is supported by suspended cables which, in turn, support the heating/cooling system together with engine rooms. The impressive radius of 51.4 m makes it one of the most striking arenas in the world.

Whereas, in the previous cases, the cables were designed to assume a resistant-by-shape configuration (i.e. funicular surface), in order to cover

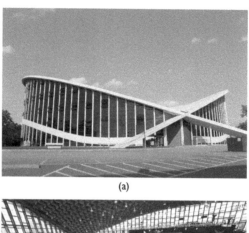

(a)

(b)

Figure 4.68 Dorton Arena in Raleigh, North Carolina (opened in 1952). (a) West side
    and (b) internal view.

large spans, the cables can also be used as tendons for concrete or steel slabs
(this technology has been widely adopted for suspended bridges since the
late 19th century). Regarding domes, the Muna Bulk Roof in Saudi Arabia
(Figure 4.70) is an example of a cable-stayed reinforced concrete roof. It is
of 340 m in diameter, covering a circular bulk storage reservoir of approximately 1 million cubic metres of water.

Figure 4.71 shows the main schemes for dome and vaults made by pneumatic and inflated structures. Regarding the former, the inner pressure allows
the external envelope (dome shaped) to stand and resist against wind and
snow loads. In particular, although it may change according to the loads, the
interior average air pressure is usually around 250 Pa higher than external
pressure (i.e. 0.25% of 1 atm), secured by airlocks at all access points. As a
consequence, the structure needs to be securely anchored to the ground and,
for wide-span structures, cables are required for anchoring and stabilization.

Inflated structures are based on tubes made by thin membrane (less
than 1 mm), which, once inflated, become 'stiff' structural elements,
which are self-contained and allow covering very large spans. The main

(a)

(b)

Figure 4.69 Madison Square Garden Arena in New York City (USA, inaugurated in 1968). (a) Construction phase and (b) internal view.

(a)   (b)

Figure 4.70 Muna Bulk Roof (Saudi Arabia, 1972). (a) Construction phase and (b) external view.

difference with respect to pneumatic structures regards the fact that in this type of structures, the pressure of the occupied interior of the building is equal to the normal atmospheric pressure. Therefore, no airlocks are needed, being anchoring and stabilization still required.

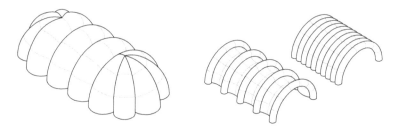

Figure 4.71 Vaults and domes: schematic views of pneumatic and inflated structures.

Finally, Figure 4.72 shows the main schemes for tension membrane structures. This type of structures is defined by elements carrying only tension and no compression or bending. Usually, (thin) shells are arranged with a double negative curvature (left side), but an external system of cables may lead to simpler shapes (right side). In both cases, the membrane structures are supported by some form of compression or bending elements, such as masts, compression rings or beams.

Following the legacy of the Russian engineer Vladimir Shukhov (1853–1939), who developed not only new structural forms (e.g. tensile, hyperboloid and gridshell structures) but also the mathematics for their analysis, since the 1960s starting with the works of the German architect and engineer Frei Otto, tensile structures have been widely adopted. Otto's first use of this concept dates back to 1967, for the Expo in Montreal (Figure 4.73a). The same idea was used by the German architect for the roof of the Olympic Stadium for the 1972 Summer Olympics in Munich (Figure 4.73b).

Nowadays, steady technological progress has increased the popularity of fabric-roofed structures. The low weight of materials may yield construction easier and cheaper than other designs, especially when vast open spaces have to be covered.

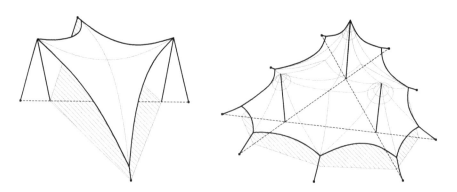

Figure 4.72 Vaults and domes: schematic views of a tension membrane structures.

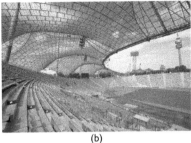

(a)   (b)

Figure 4.73 Frei Otto's main works. (a) West German pavilion at Expo 1967 in Montreal (Canada) and (b) Olympic Stadium in Munich (inaugurated in 1972).

4.3.3 Other structural systems

Before proceeding with the description of other modern structural solutions, a brief overview of the recently developed materials is provided, namely, glulam and prestressed concrete.

4.3.3.1 Glulam

Glued laminated timber, also called glulam, is a type of structural engineered wood product comprising several layers of dimensioned lumber bonded together with durable and moisture-resistant structural adhesives. Accordingly, a single large structural member is manufactured from smaller laminated pieces of lumber, shaped in different forms for elements such as vertical columns, horizontal beams or curved arch shapes. In this way, glulam provides the strength and versatility of large wood members without relying on the old growth-dependent solid-sawn timbers. This provides a more homogeneous material and also reduces the overall amount of wood used when compared with solid sawn timbers by diminishing the negative impact of knots and other defects.

Figure 4.74 shows a schematic view of the glulam production process. Although it is theoretically possible to build any type and size of elements, transportation plays a crucial role. Two examples of transportation are given in Figure 4.75.

4.3.3.2 Prestressed concrete

Prestressing a structure consists in producing a defined stress state in a controlled manner, prior to its loading, with the aim to compensate part of the stresses experienced when subjected to service conditions. In concrete, prestressing allows to reduce or fully eliminate tensile stresses, avoiding cracking and allowing a full-section stiffness for the loaded member. The technique is widespread in members subjected to bending.

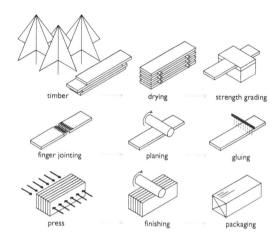

Figure 4.74 Glulam production process.

(b)

Figure 4.75 Two examples of transportation of glulam elements. (a) Variable cross section beam and (b) roof truss beam.

The concept behind prestressing is illustrated in Figure 4.76 (tension is indicated by the negative sign). The first row shows a simply supported beam with a distributed load producing a certain stress distribution (with a maximum stress σ_{max}). In the second row, the same beam is loaded by a normal action (i.e. the prestressing force) at one-third of the cross-section height. According to fundamentals of beam theory, the beam results entirely compressed with a null stress at the upper bound. For the given triangular stress distribution, it is possible to tune the magnitude of the force to achieve the requested stress. At this point, in the framework of linear elasticity, it is possible to superimpose the two stress distributions, as shown in the third row of Figure 4.76. Although the maximum compressive stress does not change, the maximum tensile stress is drastically reduced.

In order to provide the tensioning force, two main strategies are available: pre-tensioning and post-tensioning. Pre-tensioning consists in stressing the tendons before the concrete being cast. The concrete bonds to the tendons as it cures, following which the end anchoring of the tendons is released, and the tendon tensile forces are transferred to the concrete as compression through the bond between concrete and prestressing steel. For tensioning the tendons, pre-tensioning requests very stiff moulds or specifically designed grade slabs. The main steps are briefly summarized in Figure 4.77.

Differently, in post-tensioning, the tendons are tensioned after the surrounding concrete structure has been cast. The tendons are not placed in direct contact with the concrete but are encapsulated within a protective sleeve or duct. Once the concrete has been cast and set, the tendons are tensioned by pulling the tendon ends through the anchorages while pressing against the concrete, using a jack. Once the tendon is 'locked off' at

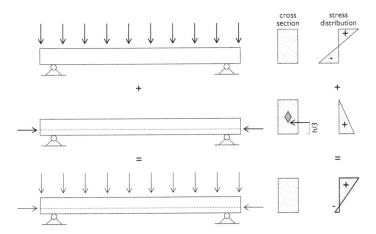

Figure 4.76 Prestressing of a simply supported reinforced concrete beam subjected to uniformly distributed load (tensile stresses are negative, and compressive stresses are positive).

Pre-stressing the tendons by
means of hydraulic jacks.
Placing passive
reinforcement and moulds

Placing concrete
Waiting until concrete can
resist prestressing

Release the force
to the members

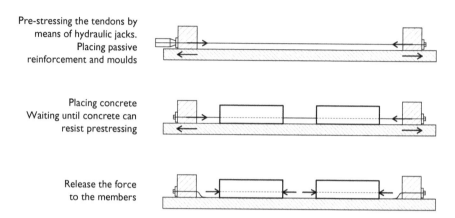

Figure 4.77 Prestressing by means of pre-tensioning tendons.

the anchorage, its tensile stress becomes a permanent compression for the concrete. The protection against corrosion may be guaranteed by cement grouting empty ducts or by grease or wax present in the duct before post-tensioning.

Figure 4.78 presents the main scheme of a post-tensioned beam together with the terminology and images of the main components of the system. It is worth noticing how, by shaping the inner duct, it is possible to locate the prestressing action in different heights within the cross section. Intuitively, as the tendon is meant to reduce the tensile stresses, it approaches the side of the beam where the tensile stress is expected, for example in the upper part of the section upon the supports or the lower part of the section at the mid-span of the beam.

The advantages of prestressing are numerous. It allows making the most from high-strength materials (the typical yield strength of prestressing steel is 1,400–1,600 MPa vs. 400–500 MPa of a steel rebar). Moreover,

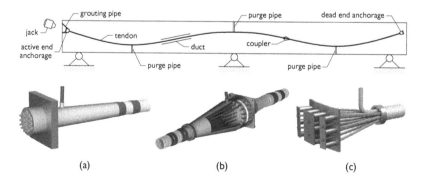

(a) (b) (c)

Figure 4.78 Prestressing by means of post-tensioning tendons: nomenclature and main elements. (a) Active end anchorage, (b) coupler and (c) dead end anchorage.

the absence of concrete cracking provides high stiffness to bending members, thus lower height-to-span ratios, providing also better durability and water tightness. The resulting performance allows extending the applications of concrete structures to loads and spans usually available to steel structures, for example in bridge engineering. On the other hand, one of the main drawbacks of prestressing is that the prestressing force decreases due to instant and delayed losses. Instant losses are due to friction, pull-in of wedges (jaws) end elastic compression of concrete. Delayed losses are due to concrete creep and shrinkage, and steel relaxation. As the performance of the structural elements depends on prestressing, these losses are duly taken into consideration in the design process.

4.3.3.3 Trusses

A truss is a structure composed of linear members organized so that the assemblage as a whole behaves as a single object. It typically comprises triangular units constructed with straight members whose ends are connected at joints (i.e. nodes) with pinned ends. In this context, external forces and reactions to those forces are considered to act only at the nodes, and the resulting action in the members is either tensile or compressive.

Besides the examples shown in Sections 3.2.3 and 3.2.4, where timber roofs and bridges were examined, respectively, Figure 4.79 shows the main schemes adopted nowadays for three-dimensional systems.

Considering precast reinforced concrete, trusses have been used, for example, for building roofs (Figure 4.80). Depending on the size of members (diagonals, struts and ties), partially assembled parts could be joined on site or each member had to be joined on site. Due to the large amount of manpower requested, in Europe, the solution was used until the 1970s.

As far as glulam is concerned, although the first application dates back to the 1840s, one of the earliest still-standing glulam roof structures is generally acknowledged to be the assembly room of King Edward VI College, a school in Bugle Street, Southampton, England, built around

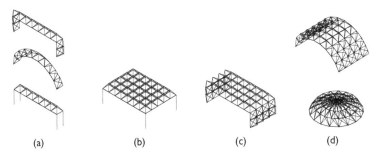

(a)   (b)   (c)   (d)

Figure 4.79 Truss: main schemes of three-dimensional systems. (a) Linear systems (used also in bridges), (b) plane systems, (c) folded systems and (d) curved systems: single and double curvature.

Figure 4.80 Roof made of precast reinforced concrete truss structure.

1866. Nowadays, the technology allows achieving high strength and stiff beams, as well as arches to span large distances without intermediate columns, with much design flexibility. Two examples are shown in Figure 4.81. The first picture shows the covered pedestrian Simme Bridge (Switzerland) built in 1989, which was designed by Julius Natterer with three spans 27–54–27 m long. The second picture shows the internal view of the roof of the Indoor Arena of Civitanova Marche (Italy), built between 2015 and 2016, with a free span of 51 m.

Double-curvature systems with variable curvature require a large number of members of different length. Its construction is supported today by computer-based design and production system. For example, the Palau Sant Jordi in Barcelona, designed by Arata Isozaki and built in 1990 for the Olympic Games (with a capacity of up to 15,000 people), is shown in Figure 4.82.

4.3.3.4 Post and lintel construction

Although the most intuitive and ancient structural system, post and lintel construction has been widely used in the 20th century and remains in use today. It is a building system where a strong horizontal element (lintel) is held up by vertical elements (posts) with large spaces between them. As shown in Figure 4.83, lintels are mainly subjected to flexure and shear forces, whereas columns are subjected to compression with a certain amount of bending (due to the force eccentricity), when gravitational forces are applied.

(a)

(b)

Figure 4.81 Two examples of glulam truss. (a) Variable cross section beam and (b) roof truss beam.

Figure 4.82 Internal and external views of the Palau Sant Jordi in Barcelona, designed by Arata Isozaki and built in 1990.

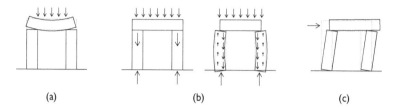

(a)   (b)   (c)

Figure 4.83 Post and lintel construction. (a) Lintel subjected to shear and bending, (b) stress distribution in the posts and (c) mechanism in case of horizontal actions.

The overall behaviour when subjected to horizontal forces is determined by the equilibrium of the columns (subjected to overturning moment).

Figure 4.84 recaps the main solutions adopted to build post and lintel structures along the history, highlighting the materials used. Until the 20th century, this represented the main solution for buildings: masonry walls with timber or steel beams belong to such type, with bracing normally provided by transverse walls. During the second half of the 19th century until the 1930s, many buildings were supported on cast iron columns, which allow rather open spaces. In this case, bracing was usually provided by transverse walls or wide load-bearing walls. After this period, mixed buildings with a masonry envelope and a concrete frame became popular, in a transition to the full frame system (see the next subsection).

On the other hand, Figure 4.85 shows the main scheme adopted nowadays, often using reinforced and prestressed concrete. In order to enhance

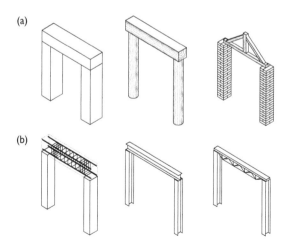

Figure 4.84 Post and lintel along the history: structural solutions and materials. (a) **Lintels can be of**: stone, wooden beams, trusses reinforced concrete, steel beams and trusses and (b) **columns can be of**: stone, stone and brick masonry, concrete, wood, steel and cast iron.

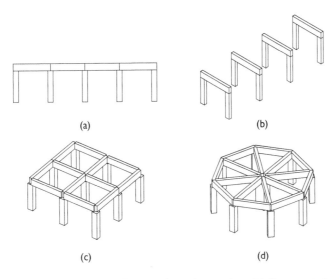

(a)

(b)

(c)

(d)

Figure 4.85 Post and lintel: main schemes adopted nowadays. (a) One way in longitudinal series, (b) one way in parallel series, (c) two-way grid and (d) circular.

Figure 4.86 Post and lintel: example of an industrial roof built with prestressed beams.

the potential of this system, prestressed concrete lintels are widely used, for example, for roofs of industrial buildings (Figure 4.86), or in precast decks for bridges.

4.3.3.5 Frames

Frames are the most common structural system of the 20th and 21st centuries. Conversely to the post and lintel solution, frame structures provide rigid connections between columns and lintels allowing an integral (global) behaviour. In this case, all the elements are subjected to bending, and the monolithic behaviour allows them to perform more efficiently. As it is possible to notice in Figure 4.87, post and lintel structures exhibit a limited

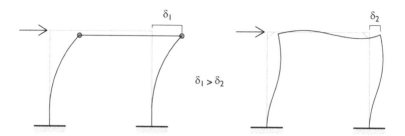

Figure 4.87 Frame structures: effect of the rigid connections between lintel and columns.

strength when subjected to horizontal forces, which have to be resisted by columns only. On the other hand, in frame structures, the horizontal element collaborates with the columns resulting in an overall stiffer structure, both for horizontal displacement and vertical displacement of the beam (here, lintel). In addition, frames resist larger vertical loads on the beam, and the mechanical slenderness of the columns presents a significant reduction.

Paralleling the space arrangement for post and lintel construction, the main schemes for frame structures are shown in Figure 4.88. In order to guarantee rigid connections at the nodes, fewer materials are adequate to obtain frames, namely, reinforced concrete, steel and wood, whereas the latter two materials require particular care in connections. Examples are given in Figures 4.89–4.92.

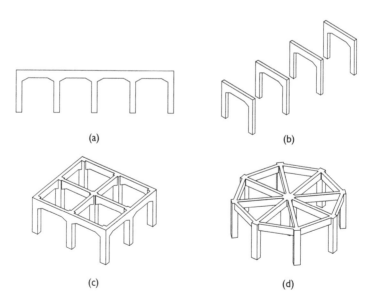

Figure 4.88 Frame structures: main schemes. (a) One way in longitudinal series, (b) one way in parallel series, (c) two-way grid and (d) circular.

(a) (b)

Figure 4.89 Reinforced concrete frame structures. (a) The first (acknowledged) building of this kind – Immeuble Hennebique (1900) and (b) typical structure today.

(a) (b)

Figure 4.90 Large-span reinforced concrete frame structures. (a) Grande Arche de La Défense, designed by arch. Otto von Spreckelsen and completed in 1989 and (b) box girder bridges.

Figure 4.91 Steel frame structures: Empire State Building during construction (1931).

Figure 4.92 Glulam frame structures: Portland Building (School of Architecture), University of Portsmouth (UK).

Chapter 5

Ancient sizing rules and limit analysis of masonry arches

Despite the modern definitions of engineers and architects, the main protagonists of the constructions of the past were masons and builders (in ancient Greece they were called *architekton*, i.e. the master builder). However, now as then, the goal of builders has been that of conceiving and guaranteeing a series of requirements within the construction. The only difference along the centuries regards the methodology applied for achieving this objective, which is the topic of the present chapter for the case of masonry arches.

Without claiming to fully treat this subject, for which the reader is referred to specific literature (e.g. Heyman, 1972a; Benvenuto, 1991; Becchi and Foce, 2002; Huerta, 2004; Addis, 2007; Kurrer, 2008), a brief overview of the evolution of structural design of the masonry arch along the history is presented. In particular, this chapter is limited to the main outcomes achieved up to the 19th century, when the masonry arch was the key protagonist. This expertise, made of rules of thumb based on experience or empirical criteria, represents nowadays an extraordinary source of information about the structural decisions made by the ancient builders for sizing vaulted structures. According to (ICOMOS/ISCARSAH, 2003), this knowledge should be used in conjunction with more modern and sophisticated approaches to assess historical masonry structures.

Following ancient achievements, a more recent approach is limit analysis, which is an easy but powerful tool to investigate masonry buildings. With the aim of providing a valuable support for practical applications, modern understanding of limit analysis is reviewed with examples of the ultimate behaviour of masonry arched structures.

5.1 HISTORICAL APPROACHES

5.1.1 Antiquity and geometrical proportions

The origin of masonry vaulted structures has been addressed in Chapter 4. Besides the religious and political symbolism, arched structures convey a sense of deep admiration and surprise. In the words of Tosca (1651–1723), '*the same gravity and weight which should have precipitate them* [the voussoirs] *to the earth, maintain them constantly in the air*' (Tosca, 1707). The price to pay for this wonder (arch as a shape-resistant structure) is the thrust which becomes evident in the overturning moment of abutments. To overcome this structural challenge, the weight has represented for centuries the only expedient, that is, buttressing by loading.

Despite the lack of scientific background, except for the simple machines studied by Archimedes (287–212 BC) first, and by Heron of Alexandria (1st century AD) then, ancient scholars achieved an extraordinary level of beauty and precision. In the words of Benvenuto (1991), '*ancient techniques slowly arrived at satisfying levels of complexity and perfection long before theory caught up with them*'. The only supports were experience and early empirical criteria, that is, a sort of 'a rudimentary scientific approach' based on the trial-and-error method, where each building could be considered a scaled specimen for a new one to be built. Based on successful achievements, ancient builders gathered competence under the so-called *rules of thumb*.

According to the classical idea of beauty based on numerical proportions, until the end of the 18th century, building rules were made up by simple geometrical definitions with notable results. In fact, as long as strength (and dynamics) is not involved, the theory of proportions provides correct outcomes (Di Pasquale, 1996). Traditionally, the base of this approach was the Pythagorean concept of beauty, that is, *concinnitas universarum partium* (the parts of a building were to be in harmony and to agree with the whole), but it is not clear how structural reasoning was introduced.

One of the first examples of this approach is discussed in the Bible, Chapter 40 of the Book of Ezekiel (ca. 597 BC): *The Man with a Measure*. The rules presented there do not regard vaulted structures, but they stress the concept of proportion, according to which changing the geometrical scale factor of a certain building does not affect its safety and capacity. On closer inspection, this is the key point of almost all the practical rules of the pre-scientific period, that is, until Christopher Wren (1632–1723) who, at the end of the 17th century, tried to apply for the first time scientific rules to the geometric tradition. For instance, although the large difference in terms of span, three Italian domes are drawn with the same height in Figure 5.1. As it is possible to notice, all of them have the same proportion between the elements.

During the Roman Empire, Marcus Vitruvius Pollio, commonly known as Vitruvius (born ca. 80–70 BC, died after ca. 15 BC), wrote the first genuine treatise in the architectural history, nowadays the only surviving major book on architecture and construction of the classical antiquity. His Ten Books on Architecture, written in Latin and Greek, provide precious information on Roman construction technology, materials and design, but no hints are provided for rules or criteria for structural design. In turn,

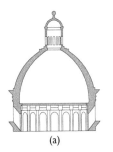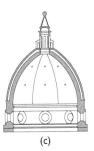

(a)   (b)   (c)

Figure 5.1 Equal proportion for three domes, after Huerta (2004). (a) San Biagio (Montepulciano), 14 m span; (b) Saint Peter (Rome), 42 m span; and (c) Santa Maria del Fiore (Florence), 42 m span.

Heron of Alexandria wrote (ca. 60 AD) about the technical aspects related to vault construction, later cited by Anthemius of Tralles (474–534), one of the architects of Hagia Sophia in Constantinople. Unfortunately, both documents were lost (Huerta, 2004).

5.1.2 Medieval ages: Gothic rules of thumb

At the end of the 5th century AD, the decline and subsequent fall of the Roman Empire led to the Early Middle Ages, characterized by an overall decline of the building yard, both in terms of techniques and materials. It is only after the 10th century that high and wide spanned vaulted structures reappeared in Central Europe, reaching the climax two centuries later with the beginning of Gothic, as discussed in Chapter 4.

Although Gothic gave a renovated impulse to architecture, mainly boosted by the introduction (probably from Middle East) of the pointed arch, no treatises of architecture regarding vault design appeared until the 15th century. During almost the entire Gothic period (12th–16th centuries), indeed, the rules were mostly handed over in secrecy, and treatises were written only after Renaissance and Baroque, with a delay of almost four centuries. It is worth stressing the importance of the *Expertise*, that is, a conference of outside masters held in case of failures or when building new important constructions. The most famous ones are those related to the Milan Cathedral (1392 and 1400), the largest church in Italy, and the Cathedral of Saint Mary of Girona (1386–1416) with the largest Gothic cross vault known, 23 m span and 35 m height (Huerta, 2004).

The only exception is represented by Villard de Honnecourt (13th century), whose portfolio (well known for the human and animal figures presented) contains 33 parchment pages with 250 drawings, including architectural details of contemporary important buildings. Although his contribution is still the oldest reference for Gothic Architecture, doubts have been placed about him not being an architect or a mason, but just a layman traveller recording what he saw. The fact that he did not reveal any truly technical and structural insight, and that he got wrong the position of the flying buttresses in the drawing of Reims Cathedral, confirms this hypothesis. As it is possible to see in Figure 5.2, as a general rule, the flying buttress should be placed between one half and one-third of the vault rise, and not at the same level of springings of the ribs. Note that the vault is filled usually up to half or two-thirds of the height, as addressed later, and there is an arch in the transversal direction to the sections shown.

One of the main concerns of the Gothic masons regarded the assessment of the adequate dimension (width) of the buttress for a given arch. In this regard, the most famous rule of thumb was the so-called *Blondel's rule*, also known as 'Fr. Derand's rule' (Derand, 1643; Blondel, 1675). It consisted in the division of the intrados of the arch in three equal parts, from which it is possible to geometrically obtain the width of the abutment (Figure 5.3a). According to Müller (1990), the rule was already cited in Boccojani's lost treatise of 1546, which means that it was defined at least during late Gothic. Despite the consistency with Gothic structures (Figure 5.3b,c), there is no evidence to consider it as a genuine Gothic rule. However, the evident utility of the rule, together with the correct ability of providing wider supports for larger thrust (when changing an arch from pointed to flat), made this rule to rapidly spread all over Europe, as mentioned by several scholars of the subsequent centuries. For instance, a similar procedure was mentioned by Martínez de Aranda (Spain, 16th century). In any case, no consideration about the height of the buttresses was made, even if this is important to stabilize them.

Along the centuries, other rules appeared, probably related to the Gothic architecture legacy. Slightly different from Fr. Derand's rule, in 1560, Hernán Ruiz el Joven introduced the arch thickness into the geometrical

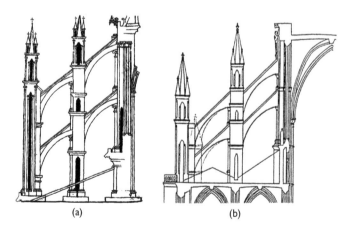

(a)   (b)

Figure 5.2 Flying buttresses in the Reims Cathedral. (a) Villard's drawing (of the apse) and (b) comparative elevation.

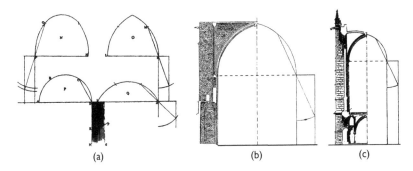

Figure 5.3 Fr. Derand's rule. (a) Application to different types of arch (Derand, 1643); (b) to the Cathedral of Girona, Spain; and (c) to the Sainte Chapelle, Paris, France (Huerta, 2004).

construction for the abutment design. Moreover, for the first time, the stabilizing importance of the infill was stressed, and it was recommended to add it until half of the arch rise, while the thickness of the arch should not be less than one-tenth of the span. The most striking examples are the German Gothic builders, as they set up a list of geometrical proportions that, without any clearly identified structural purpose, starting from the span of chorus, led up to the smallest details, for example the vault ribs cross section, as reported by Coenen (1990).

During the 15th and 16th centuries, during the Late Gothic, two main architects provided significant improvements in Spain, namely Rodrigo Gil de Hontañón and Friar Lorenzo de San Nicolás. The former proposed geometrical proportions and analytical formulations for designing up to minor elements. The latter addressed all aspects of constructions, becoming one of the key persons of the time. His contributions regarded the design of one-nave churches with barrel vault and lunettes, bridges and towers. For further comments about ancient design rules for cross vaults, the reader is referred to Gaetani et al. (2016).

Nonetheless, it is in Italy that Renaissance paved the way for a systematic schematization of building rules. Starting from the lessons of classical architecture, first and foremost, Vitruvius' treatise, artists and architects like Alberti, Palladio or Fontana provided rules for designing different structural members (arches, buttresses, bridges, domes or towers). In particular, Alberti's *De Re Aedificatoria* (1435-40-1452), subdivided into ten books, review the construction knowledge up to the 15th century. Translated into several languages, this was the classic treatise on architecture (the 'official doctrine' on construction) up to the mid-18th century. For instance, Figure 5.4 illustrates Alberti's rules for bridge design in a restitution of Straub (1992). As discussed for the Gothic, the rules were normally set as geometrical constructions or proportions between different structural parts.

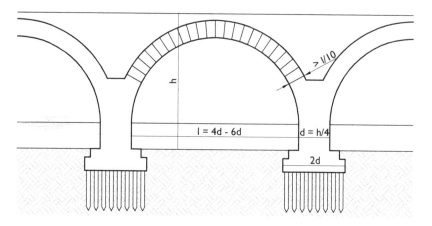

Figure 5.4 Alberti's rules for the design of bridges with circular shape arches (Straub, 1992).

5.1.3 Eighteenth century: from catenary to collapse mechanism

By the mid-17th century, scientists and designers experienced the need for rational principles or methods stemming from scientific theories. This necessity was clearly stressed by Gautier (1660–1737) in 1716, '*Only by trial and error, it was possible to build all bridges and vaults of all the buildings. Specific rules which allowed to know the safety capacity of these works have never been followed*' (Huerta, 2004). However, the new scientific approach was not yet fully autonomous and mature, meaning that, in this context, the rules of thumb still played a fundamental role. Validated by centuries-old history, the traditional rules were the only support to confirm new theories (Benvenuto, 1991; Kurrer, 2008).

Leonardo da Vinci's '*An arch is nothing but a strength caused by two weaknesses*' (ca. 1500), although in his typical cryptic style, represents the first scientific hint toward the understanding of the masonry arch. Considering the arch composed by two quarters of circles, which are not able to stand independently (weakness), they become stronger if joined. Leonardo probably realized how, due to the shape of the arch, the gravitational load becomes thrust that pushes the two curves together, reinforcing their unity. Moreover, Leonardo's sketches within Foster Codex provide suggestions, hints and concepts, which were to be developed only three centuries later. Later on, Baldi and Fr. Honoré Fabri, in 1621 and 1667, respectively, were probably the first who wrote scientific documents on the statics of the arch, the latter approaching the solution by a three-hinged arch (Benvenuto, 1991; Kurrer, 2008).

Despite these first studies, the principle of the inverted catenary (arch-catenary analogy) stated by Robert Hooke (1635–1703) in 1675 is commonly adopted as the first scientific contribution regarding the structural understanding of masonry arch. The scholar provided the solution by means of an anagram deciphered only after his death: '*Ut pendet continuum flexile, sic stabit contiguum rigidum inversum*,' that is, as hangs the flexible line (also called funicular, from Latin *funiculus* that means 'cord,' 'rope'), so but inverted will stand the rigid arch. Hooke did not provide any proof for his intuition, independently extended by Gregory (1659–1708) in 1690, and also discussed by Jakob Bernoulli (1654–1705) in 1704, who finally indicated the catenary as the most appropriate shape for a compression-only thin masonry arch with dry joints. Still, Gregory refined the concept of catenary in 1698 as a stability condition: an arch of any shape is stable only if it is possible to fit an inverted catenary within its thickness (so called anti-funicular). Obviously, every geometry and load distribution lead to different catenaries (see Section 5.2.1).

Regardless of the scientific efforts, it is possible that the first application of the catenary for determining the shape of an arched structure might date back to almost 150 years before. Attributed to Bartolomeo Ammannati (1511–1592) and probably influenced by a very old Michelangelo, the bridge Santa Trinità in Florence was apparently designed as a suspended chain (surprisingly) turned through 90° (Figure 5.5). It was built starting from 1567 and destroyed by German bombs in 1944. The study of Riccardo Gizdulichto, oriented to rebuild it 'where and how it was', highlighted the analogy with the catenary form. It was inaugurated in 1958 (Kurrer, 2008).

If the previous scholars were interested in the force distribution, de La Hire (whose theory was embraced by Belidor with a few improvements) addressed the study of the arch from the kinematic point of view. As a member and professor at the Académie Royale d'Architecture, starting from 1687, de La Hire (1640–1718) tried to give a scientific explanation to the traditional rules. In particular, he assumed the arch as a block assemblage and attempted the assessment of abutment dimensions according to the following hypotheses

Figure 5.5 Comparison between Bridge Santa Trinità in Florence (1567) and the catenary form turned through 90°.

(refer to Figure 5.6): (1) the possible crack occurs at an intermediate section between springings and keystone (45° from the arch springing, in case of a semicircle), (2) the blocks between the cracks are stable enough to guarantee a rigid body behaviour (accordingly, the foundations are perfectly steady) and (3) the thrust falls on the bottom of the breaking joint. The problem is then solved assuming that the moment produced by the central block is counterbalanced by the moment produced by the weight of the stabilizing part.

De La Hire's assumptions led to consider the central body as a wedge sliding upon the lateral blocks. This means that he assumed a null friction angle at the interfaces, which is, instead, the essential parameter for the arch to stand. Although scientifically incorrect (but important for the subsequent developments), this theory matched the traditional outcome and practice; thus, de La Hire's rule swiftly spread over Europe together with the everlasting Fr. Derand's rule.

After de La Hire, Couplet was an eminent figure of the Age of Enlightenment. His assumptions (Couplet, 1729, 1730), namely infinite friction, infinite compressive strength and null tensile strength, remain as today's hypotheses of limit analysis for analytical assessment of masonry structures (see Section 5.2). The assumptions imply that the voussoirs can only rotate around their edges at the intrados and extrados creating hinges. Consequently, the collapse of an arch follows the development of a certain number of hinges that cause the arch to become a mechanism (in general, four-hinged rigid bodies in modern terms). This understanding was confirmed by experiments carried out by Danyzy in 1732 (Figure 5.7) and others. Regarding the location of the hinges, Couplet incorrectly assumed them placed at the arch base, at the arch centre and at 45°.

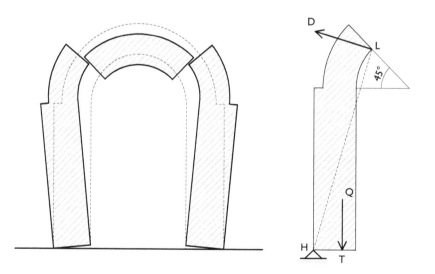

Figure 5.6 Collapse mechanism according to de La Hire (Benvenuto, 1991).

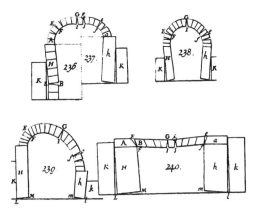

Figure 5.7 Collapse mechanism after Couplet's assumptions: outcomes of the experimental campaign performed by Danyzy in 1732.

Finally, about 30 years after Couplet, in 1773, Coulomb (1736–1806) proposed the first general and accurate theory on the stability of masonry arches. The basic assumptions are as follows: (1) sliding between voussoirs is unlikely, due to the existing frictional forces; (2) collapse is caused by the rotation between parts, due to the appearance of a number of hinges. The location of the hinges is *a priori* unknown, and the idea of considering any joint as the possible failure location allowed him to calculate an admissible range for the thrust (method of '*maxima and minima*'). The mechanism individuated by Coulomb is sketched in Figure 5.8. The study of the masonry arch was eventually finalized by Mascheroni (1750–1800) in 1785.

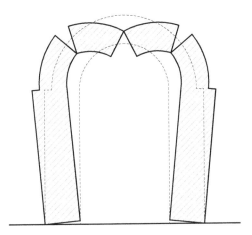

Figure 5.8 Coulomb's collapse mechanism (Benvenuto, 1991).

5.1.4 Nineteenth century: line of thrust and graphical statics

The 19th century, with Navier's *Leçons* of the 1820s (Navier, 1785–1836), is usually identified as the starting point of the elasticity theory, linking stress analysis and material strength. However, following the contributions of Hooke and Gregory about the catenary (see Section 5.1.3), several scholars were interested in the so-called *thrust line*, that is, the locus of the points through which the resultant force passes. The thrust line can be built by means of graphical statics, which is a branch of statics that deals with combination of forces (e.g. the block weights) idealized as vectors, that is, geometrical entities provided with magnitude and direction (see Section 5.2.1).

The use of vectors and parallelogram law are so intuitive that their origin is unknown, and probably, they date back to Aristotle (384–322 BC) or Heron of Alexandria (1st century AD). One of the first scholars who scientifically derived the parallelogram rule for calculating the resultant force was Stevin (1548–1620) in his book *Beghinselen des Waterwichts* (Principles on the Weight of Water, 1634). Subsequently, Varignon (1654–1722) introduced the funicular polygon and the polygon of forces in his work *Nouvelle Mécanique ou Statique*, published posthumously in 1725. Eventually, Culmann (1821–1881) discovered in 1864–1865 the structural relationship between funicular polygon and polygon of forces through a planar correlation of the projective geometry.

Going back to the thrust line, the first author who provided a complete theory was Young (1773–1829), but his work remained unnoticed until recent past (Huerta, 2010). It was only in 1831 when Gerstner (1756–1832) intuitively established the theory according to which the capacity increases with the number of indeterminacies, as the problem is statically indeterminate (Kurrer, 2008).

In this scenario, of particular interest is the contribution of Méry (1805–1866) regarding the stability of the masonry arch. Paralleling Gregory's statement, together with Navier's indications, Méry stated in 1840 that the arch is safe if it is possible to draw a thrust line inside the core of every cross section. In case of rectangular section, according to the fundamentals of beam theory, the upper and lower vertices of the core run at a distance from the extrados and intrados equal to one-third of thickness, that is, the thrust line should pass through the band defined by the middle third of the section. According to the elastic theory, this assumption guarantees no tensile stress within the element. Although the approach may lead to rather conservative results (see Section 5.2), it is still adopted by professionals and researchers to relate the position of the thrust line to cracks and hinge formation.

Despite the encouraging developments of thrust line and graphical statics, starting from the 1860s, the rise in popularity of wrought-iron structures somehow shifted the focus of the research almost exclusively upon elastic analysis, that is, on a material assumed to be (1) continuum;

(2) linear elastic, isotropic and homogeneous; and (3) deformable but with small displacements and strains. These assumptions are not accurate for describing the nonlinear behaviour of masonry, and the scholars of the late 1800 were aware of this fact. For instance, the iterative approach proposed in 1879 by Castigliano (1847–1884) is an emblematic example of the will to neglect the tensile contribution (*'masonry arches as imperfectly elastic systems'*). Moreover, almost 30 years later, Sejourné (1851–1939) stressed in 1913 the inaccuracy of elastic analysis, unwillingly accepted until a new formulation would have been able to better match the reality. In fact, limit analysis would have not delayed much.

5.2 LIMIT ANALYSIS

Although rudiments of masonry ultimate behaviour and stability conditions have been already stressed during the 18th and 19th centuries, it is only after the 1930s that a systematic approach was devoted to plasticity and limit analysis. Research aimed at better investigating the reserve of strength exhibited by steel members when compared with models based on the elastic theory. The evident reasoning was that a more accurate theory would provide a saving of economic resources. In this scenario, limit analysis played a prominent role for getting quick and valuable information about the ultimate behaviour of structural elements.

The Soviet engineer Gvozdev (1897–1986) was the first who provided and verified the fundamental theorems of ultimate load design. However, his work (published in 1938) was only available in Russian and through the Moscow Academy of Sciences, whose publications were little known in the West. Gvozdev's outcomes were 're-discovered' by the Anglo-American research group led by Prager (1903–1980) and others. Baker (1901–1985), Horne (1921–2000) and Heyman (1925–), in their book *The Steel Skeleton. Vol. 2: Plastic Behaviour and Design* published in 1956, provided the first work to mention the fundamental theorems of ultimate load theory and describe practical applications.

Heyman was also the first who recognized the possibility of adopting limit analysis for other materials with a ductile behaviour, provided that there are no instability problems. If for reinforced concrete these assumptions can be accepted under many circumstances, it was still not clear how to deal with timber and masonry. Regarding the latter, with particular reference to the masonry arch, following Drucker's (1918–2001) early attempts, Kooharian published the first modern work on the subject in 1952. Eventually, 'The stone skeleton' (Heyman, 1966) explained the hypotheses for performing limit analyses of any masonry load-bearing structure: (1) infinite compressive strength, (2) zero tensile strength and (3) absence of sliding failure. On closer inspection, these assumptions are the ones proposed by Couplet more than two centuries before. Also Heyman (1982) stated that Couplet's analysis represented the historic-logical starting point of ultimate load theory approach to masonry arch statics.

More in detail, the first hypothesis is reasonable, because masonry tends to resist well to compressive actions, and historical buildings usually display stresses under service loading that are one or two orders of magnitude below the crushing strength of stone (Table 5.1). Here, it is stressed that the compressive strength of masonry is usually much lower than that of stone. The second hypothesis may be slightly conservative because the possible presence of mortar at the joints usually provides some weak adhesion (typically in the range of 0.05–0.20 N/mm²), eventually vanishing to zero under progressive rotation. Finally, the third assumption assumes an infinite friction coefficient. Although this is not correct (real values are usually in the range of 0.60–0.75, which corresponds to an angle of friction of about 35°), sliding failures are quite rare in slender masonry arches, and the assumption can be considered adequate in most cases (the analyst can easily verify it).

Limit analysis does not require much computational requirements and knowledge of the practitioner (e.g. material properties, competence regarding the analysis tools and interpretation of results) but it is still powerful, if compared with the more sophisticated nonlinear analyses that are widespread today in modern engineering. Figure 5.9 shows a comparison between limit analysis and nonlinear analysis (allowing cracking but with zero tensile strength) in terms of capacity curve for an arch loaded at quarter span (load factor vs. vertical displacement). As it is possible to see, the capacity curve is able to track the entire behaviour of the structure from linear elasticity until the maximum load capacity and beyond it, providing the same capacity as limit analysis, if nonlinear geometrical effects (i.e. large displacements) are neglected. If nonlinear geometrical effects are considered, the capacity of the arch is reduced to a moderate extent. On the other hand, limit analysis provides only maximum capacity with no information about the load history, assumed proportionally increasing with respect to a scalar, that is, the load factor (or multiplier) that scales the value of the point load.

Table 5.1 Stresses under service loading in some of the largest masonry buildings

Stress under service loading	N/mm²
Column of Toussaint d'Angers Church	4.4
Column of Mallora Cathedral	2.2
Column of Hagia Sophia in Constantinople	2.2
Column of Beauvais Cathedral	1.3
Base of Pantheon in Rome	0.6

Source: Huerta (2004).

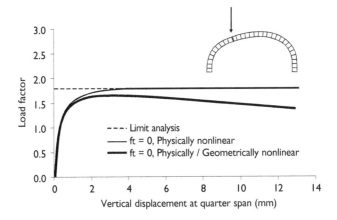

Figure 5.9 Comparison between limit and nonlinear analysis for a masonry arch loaded at quarter span with point load (Lourenço, 2001).

According to Heyman's assumptions, the goal of the analysis shifts from strength to stability, which is a matter of geometry and independent from scale. The number of parameters required for limit analysis is also low, representing an advantageous aspect given the difficulty in obtaining reliable mechanical data for historical masonry structures. The consequent results represent an accurate estimation of the ultimate capacity of a structure, provided that there are no large movements of supports, and sliding failure can be neglected. Without discussing the theoretical bases of limit analysis, for which the reader is referred to Heyman (1982), the assumptions lead to three theorems:

1. **Static or lower bound theorem**: the load λ_F for which the stress state satisfies equilibrium and yield condition is less than (or at most equal to) the true value of the collapse load. Accordingly, the theorem is referred to as 'safe' (if not exact, any result is on the safe side), and the true load factor is determined by searching for the maximum value of λ_L. For the sake of clarity, a yield condition (or yield criterion) describes the material failure in structural plasticity, that is, it represents the threshold state of the material between elastic and plastic (or brittle) deformation. It is commonly visualized as a two-dimensional surface in the principle stress space. It is possible to define three behaviours associate to the position of the stress state: (1) elastic, inside the yield surface; (2) plastic, on the surface; and (3) inadmissible, outside of the surface.

2. **Kinematic or upper bound theorem**: the load obtained from the work equation written for any kinematic admissible mechanism is greater (or at most equal to) the true collapse load. Accordingly, the theorem

is referred to as 'unsafe' (if not exact, any result is not on the safe side), and the true load factor is determined by searching for the minimum value of λ_U. For the sake of clarity, a kinematic admissible mechanism is a distribution of displacement increments consistently related to the plastic strain increments and boundary conditions.

3. **Uniqueness theorem:** if the three conditions, namely, equilibrium, mechanism and yield, are satisfied, then the load factor obtained from the static and kinematic approach is the same and is equal to the true load factor λ_F. In other words, the mechanism is the real ultimate one, and the stress state is the only possible one. In this case, the maximum load factor of the static approach and the minimum load factor of the kinematic approach are the same (Figure 5.10).

Regarding the static theorem, it is important to stress the fact that once a statically admissible equilibrium state is found, the structure is safe, no matter if this state is the actual one. This statement was already made by Hooke and extended by Gregory almost 200 years before: the arch is stable if at least one of the infinite equilibrated thrust line falls entirely into the arch thickness. This gives also a theoretical foundation for the graphical method, which, in the late 19th century, became the most used and popular approach for historical constructions.

5.2.1 Graphical statics and static approach

5.2.1.1 Introduction

Before providing practical examples for limit analysis applications, it is important to define three main concepts related to the addition of vectors (a vector is a geometrical entity provided with magnitude and direction to represent a force, which can be freely translated along its direction):

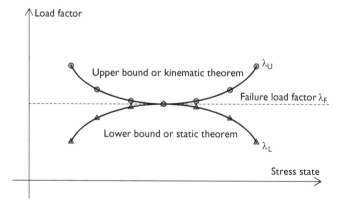

Figure 5.10 Conceptual representation of limit analysis theorems.

- **Parallelogram method:** in order to add the two vectors shown in Figure 5.11a, they must be translated to a common origin (the inter-section point **O**), and a parallelogram needs to be constructed on them as two pairs of parallel sides (Figure 5.11b). The resultant vector is the diagonal of the parallelogram drawn from the common origin.
- **Polygon of forces:** two vectors **A** and **B** can be added by drawing the vectors in such a way that the initial point of **B** is on the terminal point of **A**. The resultant **R** = **A** + **B** is the vector from the initial point of **A** to the terminal point of **B**. Many vectors can be added together in this way (Figure 5.12a) by drawing the successive vec-tors in a head-to-tail fashion, as shown in Figure 5.12b. As usual, in additive operations, the result does not depend on the adding order (or sequence). If the polygon is closed, the resultant is a vector of zero magnitude and no direction is defined (null vector or point). Considering vectors as forces, a null resultant means that, if the forces are applied to the rigid body or geometrical point, the system is balanced (in equilibrium).
- **Funicular polygon:** it is a graphical construction used to calculate the application point of the resultant of a system of forces. Looking at Figure 5.13, in order to construct the funicular polygon, initial and terminal points of each vector of the polygon of forces must be con-nected to an arbitrarily located pole **O**, as lines 1–5 in Figure 5.13b. The directions starting from **O** must be added to the system of forces in the fashion of Figure 5.13a: the intersection of the first and the last directions (here, lines 1 and 5) is the point through which the resul-tant force passes.

The funicular polygon represents also the shape of an ideal weightless thread fixed at the ends and tensioned by the forces acting on it. In this regard, each triangle with vertex **O** in Figure 5.13b can be seen as a sys-tem of balanced forces. For instance, considering the vector **A**, the local

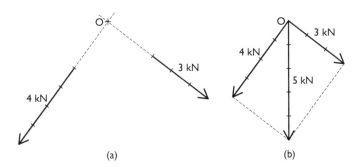

Figure 5.11 Parallelogram method for vector addiction. (a) Two vectors with their common origin and (b) parallelogram construction and resultant.

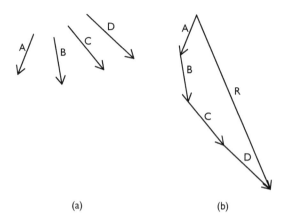

Figure 5.12 Polygon of force. (a) Four general vectors and (b) obtained resultant from addition.

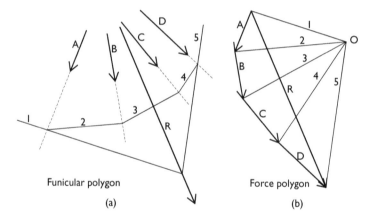

Figure 5.13 Relation between the (a) funicular polygon and (b) force polygon.

equilibrium of the thread on the application point of **A** is guaranteed with the axial forces acting along directions 1 and 2. Given the null flexural stiffness of the thread, these directions represent the shape of the thread close to the application point of **A**. Analogously, the case of parallel vectors is illustrated in Figure 5.14.

5.2.1.2 Masonry arch

Regarding the analysis of masonry arch under gravitational loads, according to Heyman's assumptions, the hypotheses of the static theorem are implicitly satisfied if the resultant force (in equilibrium with the external loads) is totally located inside of the geometry of the structure. In an

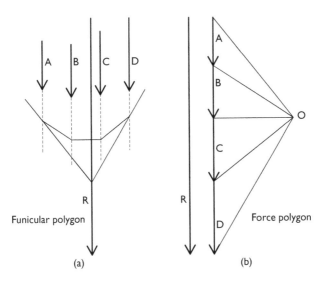

Figure 5.14 Parallel vectors. (a) Funicular polygon and (b) force polygon.

analogy with the funicular polygon, which represents the shape of a thread in equilibrium with external loads (i.e. the locus of tensile-only stresses), the shape of a compression-only system is basically the anti-funicular, that is, the inverted polygon (i.e. thrust line or the locus of compressive-only stresses). Nevertheless, since the arch is a three-time statically indeterminate structure (when assumed clamped at the two supports), several thrust lines can be found according to three different quantities. Generally, these quantities are value, direction and application point within the height of the cross section of the reaction on one side of the arch.

In order to apply the static theorem, the arch must be decomposed into a series of real or fictitious voussoirs separated by planes not necessarily parallel. For each voussoir, its weight should be applied in the centre of mass, as shown in Figure 5.15, together with possible additional load (e.g. arch fill). The vectors representing the vertical load of each voussoir should be sorted in the polygon of forces as shown. As already stressed, the location of the pole O is arbitrary, that is, its horizontal and vertical coordinates are not fixed. These two variables are associated to two of the three indeterminacies, that is, value and direction of the reaction on one side of the arch (in particular, the horizontal distance of O from the weight vectors represents the horizontal component of the thrust). The last indeterminacy regards the application point of one of the external reactions, that is, the anti-funicular can be translated horizontally. However, in case of symmetry of both geometry and load distribution, the problem loses one indeterminacy, that is, the thrust line and the funicular polygon must be symmetric too. In this case, it is possible to study half of the arch, knowing that the thrust line at the keystone (at the intersection with the symmetry axis) is horizontal.

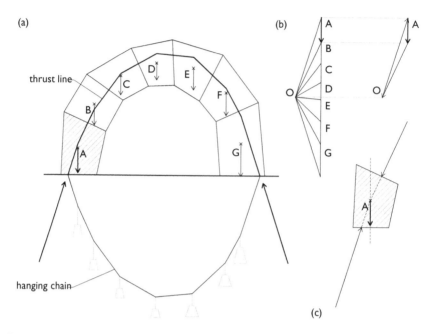

Figure 5.15 Random arched structure after Block, DeJong and Ochsendorf (2006), a possible thrust line and its equivalent hanging chain (funicular polygon) (a), together with the relative polygon of forces and the vector representation of the voussoir's equilibrium (b) and forces acting on the same voussoir (c).

Figure 5.15 shows the study of a random arched structure. The anti-funicular (thrust line) is totally inside of the arch thickness, thus the arch is safe and collapse will not occur. The balanced forces acting on an isolated voussoir and its relative polygon of forces are also shown.

Finally, Figure 5.16 presents the two extreme thrust-line configurations for a semicircular arch under self-weight. It is worth noting that the config-uration with the minimum horizontal thrust is achieved with the thrust line close to the extrados at the keystone, consequence of an outward support horizontal movement, see Figure 5.16a. This would be the usual behaviour of an arch with sliding of the supports due to rotation of the side walls or insufficient buttressing. Analogously, the maximum horizontal thrust is the one with the thrust line approaching the intrados at the keystone, Figure 5.16b. This configuration occurs when the arch is subjected to an inward support horizontal movement. This can be the case of a three-nave church in which the centre nave pushes inwards a strongly buttressed lat-eral nave. It is easy to imagine how several thrust lines can be built between these two configurations. For an interactive analysis of masonry structures, the reader is referred to MIT (2005). Figure 5.16 shows also that hinges occur once the thrust line touches (i.e. becomes tangent to) the boundary (see Section 5.2.2).

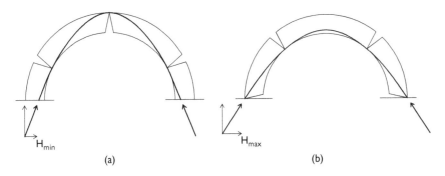

Figure 5.16 Semicircular arch under self-weight. (a) Minimum thrust and (b) maximum thrust (Heyman, 1995).

In case the arch is subjected to a vertical point load as illustrated in Figure 5.17, the thrust line can be easily calculated according to the procedure described in Figure 5.14. The external load can be simply treated as an extra vector in the polygon of forces. It is worth noticing that, according to influence line analysis, the arch exhibits the lowest capacity when the external load acts at about a quarter span. For the sake of clarity, an influence line is the graphical representation of a function at a specific point on a structure caused by a unit load placed at any point along the structure. In this case, the function is the arch capacity.

Figure 5.18 details the application of the static approach to a masonry arched portal. Given the symmetry of the problem, it is possible to study half of it by imposing the thrust line to be horizontal at the intersection with the symmetry axis (**H**). This means that the pole **O** of the polygon of forces must be placed on the horizontal axis passing through the initial point of **A**. As already stressed, thanks to symmetry, the indeterminacies are two, in this case, the value and the position of **H** within the height of the arch at the keystone.

Starting from the reaction R_1 calculated in Figure 5.18, the subsequent phases are shown in Figure 5.19, where all the other weights in the portal are given. Applying the parallelogram method recurrently (thanks to the associative property of vector addition), it is possible to calculate the final reaction

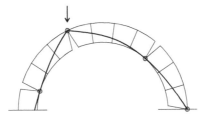

Figure 5.17 Thrust line for an arch subjected to an external point load.

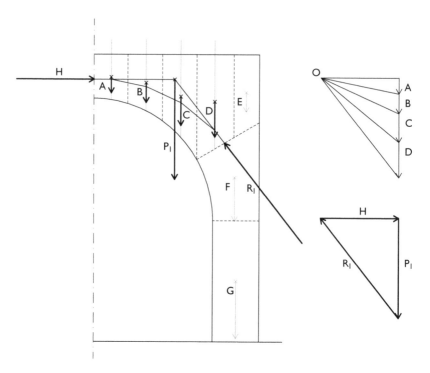

Figure 5.18 Application of the static approach to a symmetric masonry arched portal.

at the level of springings. As illustrated in Figure 5.19, the thrust line lays outside of the structure, that is, the structure is not stable, at least for the given value and position of the **H** (Figure 5.18). In case a stable configuration cannot be reached changing these two variables, it is possible to (1) add more weight on the buttress (i.e. buttressing by loading to force the thrust line to become more vertical, in the fashion of Gothic architecture), for example, by adding a pinnacle, floors or fill material to an arch and (2) increase the cross section of the buttress to contain the thrust line within its geometry. In the case of buttressing, the weight of the new part also contributes to reduce the horizontal thrust but, in order to be effective, an added buttress must be properly connected to the original structure to ensure monolithic behaviour.

Finally, to evaluate the capacity of a transversal cross section of an entire masonry building, it is possible to build the thrust line until the foundation. In this regard, Figure 5.20a shows the graphic analysis of Mallorca Cathedral performed by Rubió y Bellver (1912) by means of hand calculation, showing that the cathedral is safe for gravitational loading (assuming no soil settlements). Thanks to modern computer applications, nowadays this work is less tedious, and Figure 5.20b shows numerous possible thrust-line configurations contained within the same structure (Maynou, 2001).

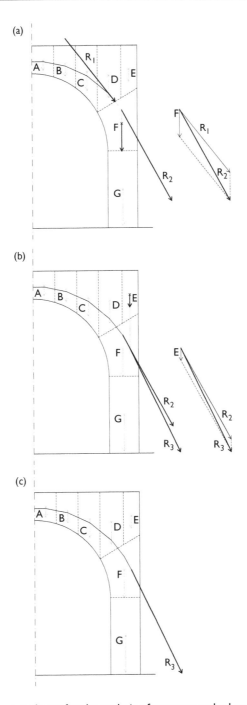

Figure 5.19 Subsequent phases for the analysis of masonry arched portal.

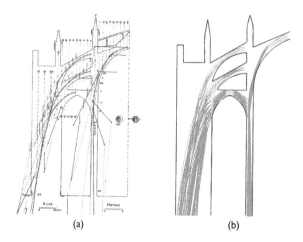

(a)   (b)

Figure 5.20 Graphic analysis of Mallorca Cathedral under gravitational loads. (a) Analysis performed by Rubió y Bellver (1912) and (b) various numerous solutions found with a computer application (Maynou, 2001).

5.2.1.3 Vaulted structures

The procedure described so far is inherent to (in-plane) sectional analysis. The application of the static approach for three-dimensional vaults is rather complicated, but hand calculations can still be performed according to the *slicing technique*. Pioneered by Frézier (1737), this method allows disassembling the vaults in elementary arches, that is, a reduction of a three-dimensional problem into an in-plane one. The first case where this method was applied regards the dome of Saint Peter's Basilica in Rome studied by Poleni (1683–1761) and described in his report of 1748 (see Section 4.2.4). Following the damage observation, he conceived the dome as an assemblage of independent angular sectors that could be studied according to the thrust-line method. Although he did some miscalculation in the preliminary study, in his second attempt, Poleni was able to calculate the anti-funicular considering the correct geometry of the fictitious voussoirs which got smaller when approaching the top of the dome (Figure 5.21).

With the modern understanding, the slicing approach has a clear structural reasoning. Considering the elastic analysis of a dome, the circumferential (or hoop) stress distribution is shown in Figure 5.22a. As it possible to verify, the upper part of the dome undergoes compressive stresses, whereas the lower part of the dome experiences tensile stresses, which are larger in magnitude. Considering that masonry possesses very low tensile strength, this means that the lower part of the dome usually exhibits radial cracks, transforming the dome in a series of almost independent sectors

(a)

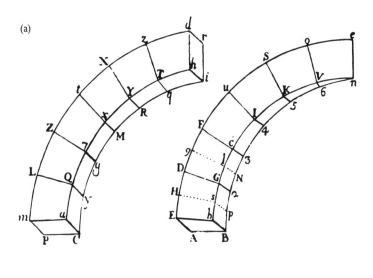

(b)

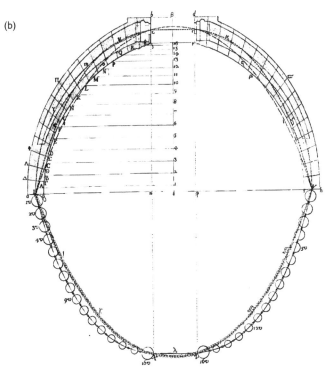

Figure 5.21 Poleni's analysis of the Saint Peter's Basilica in Rome. (a) Difference between an arch and a sector of a cupola and (b) funicular analysis of the dome (Poleni, 1748).

(Figure 5.22b). These, in fact, still interact in the upper part of the dome (note that the part subjected to compression becomes smaller with respect to Figure 5.22a, due to the crack progress and lack of tensile forces).

According to the approach proposed by Frézier and adopted by Poleni, it is also possible to analyse a cross vault under gravitational loads. The reasoning is based on intuition but the procedure is more complex. Figure 5.23 two examples of graphical analysis of cross vaults. Every web is decomposed in parallel strips studied considering the minimum thrust, and their reactions become external loads for the diagonal arch. In case of double-curvature vaults, Ungewitter and Mohrmann (1890) suggested to divide the webs in elementary arches following the idea of a ball rolling down the extrados (Figure 5.24a). The same idea was followed by Sabouret (1928) and Abraham (1934); however, since only the latter provided explicative drawings (Figure 5.24b), the entire credit was given to Abraham (Huerta, 2009).

As in the case of the dome, slicing schemes follow from existing damage observation. In particular, the well-known Sabouret cracks, which are the cracks running parallel to the side walls, typical of Gothic cathedrals, can be understood as a consequence of outward support movement (Figure 5.25). Extending the reasoning to the rest of the system, it is easy to consider the vault as a series of parallel arches. Nonetheless, more recent understanding considers these cracks as a consequence of compatibility problems in the connection between the lateral vault and wall.

In short, according to the static theorem of limit analysis, infinite admissible configurations can be drawn to assess the stability of a given structure. Alternative decompositions may be possible (although difficult to be determined), but this does not mean that all of them are equally realistic and even safe (due to the fact that soil settlements can be present, not to

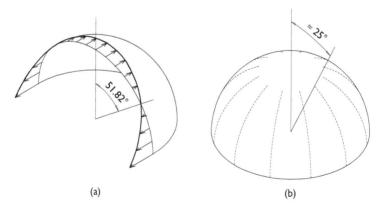

(a)   (b)

Figure 5.22 Dome analysis after Heyman (1995). (a) Circumferential (or hoop) stress distribution and (b) typical crack pattern.

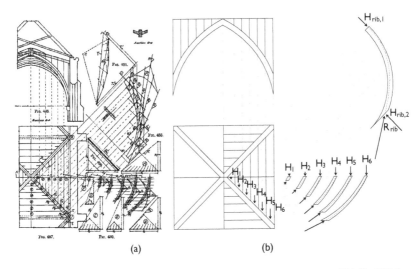

Figure 5.23 Graphical statics applied to cross vaults. (a) According to Wolfe (1921) and (b) typical case after Block (2009).

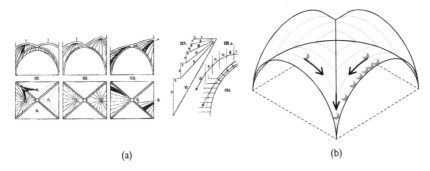

Figure 5.24 Slicing technique. (a) Patterns of slicing (Ungewitter and Mohrmann, 1890) and (b) 'ball principle' after Abraham (1934).

say other loading than gravitational effects). In general, a few concepts are worth to be stressed. Whenever subjected to tensile stresses, domes and vaults experience cracks, developing a system of compatible arches contained within the volume of the structure. If the structure cannot generate an equivalent system of compatible arches, it is likely to collapse due to the loss of tensile strength in the medium or long term (or the absence of enough tensile strength from the beginning). Since cracking is a quasi-brittle phenomenon, the ability of the structure to successfully develop a possible system of compatible arches is not warranted even if a possible system can be envisaged. Additional calculations may be useful to ascertain the true resisting mechanism.

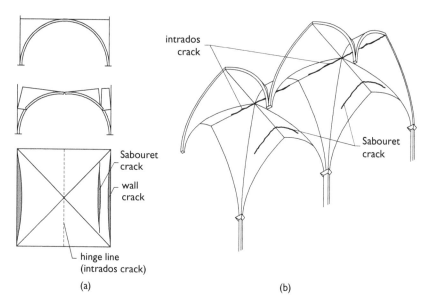

Figure 5.25 Traditional interpretation of Sabouret cracks. (a) After Heyman (1983) and (b) schematic representation, bottom view.

The slicing technique has two main drawbacks: it neglects the inter-action between two adjacent slices, for example the favourable com-pressive circumferential stresses in the dome, and it is generally based on a pre-defined force path configuration, for example cross vaults are studied as independent web strips, whose resultant action is applied to the diagonal arch. Recently, thanks to automatic procedures, more refined procedures have been proposed (O'Dwyer, 1999; D'Ayala and Casapulla, 2001; Andreu et al., 2007; Block, 2009). Figure 5.26 shows three possible alternative force patterns for a cross vault analysis under gravitational loads.

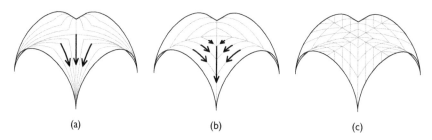

Figure 5.26 Different force distributions for a cross-vault analysis according to O'Dwyer (1999). (a) Forces towards the corners, (b) parallel to lateral arches and (c) finer pattern.

5.2.2 Kinematic approach

5.2.2.1 Introduction

The most relevant aspects of two-dimensional kinematic analysis of mechanisms are detailed next, referring the reader to Heyman (1982) for further details:

- **Mechanism:** despite the more accurate mechanical meaning, here mechanism is assumed as a *linkage* where rigid parts are interconnected so that the motion of each part is constrained to each other. The analysis of a mechanism is based on the concept that any displacement of a rigid body having motion in its plane can be considered as a pure rotational motion (of the body as a whole) about some instantaneous centre of rotation (which has in addition two translation movements). In the present study, only single degree of freedom (SDOF) mechanisms are considered, that is, the displacement of all the rigid parts can be described according to only one arbitrary displacement.
- **Instant centre of rotation (ICR):** the ICR can be referred either to a rigid body or to a mechanism, defined as 'fixed' and 'permanent', respectively. The former is the point whose displacement (both rotation and translation) is zero at a given instant. The fixed ICR is unique for each body and can be a point not belonging to it. The latter is a common point between two bodies where the displacements are equal, both in direction and magnitude, that is, it is the point around which the two bodies rotate with respect each other. Also in this case, it can be a point not belonging to either of the bodies but, conversely to fixed ICR, permanent ICR moves when the mechanism moves. Moreover, since any given body can rotate with respect to others, the number of permanent ICRs is equal to $n(n - 1)/2$, with n number of bodies. For instance, considering $n = 3$ bodies (1, 2, 3), there are $n = 3$ fixed ICRs (C_1, C_2, C_3) and 3 permanent ICRs (C_{12}, C_{13}, C_{23}).

    The first step in the study of a mechanism is the identification of ICRs by visual inspection. The external constrains give information about the fixed ICRs (Figure 5.27a): (1) a pin/hinge is a fixed ICR itself, (2) the fixed ICR for a slider is at infinite distance in a direction perpendicular to the sliding motion (i.e. a rotation with an infinite radius is a pure translation) and (3) the fixed ICR for a roller can be any point belonging to its axis.

    Regarding the permanent ICRs, they are determined according to the internal constrains between two bodies. Analogously, looking at Figure 5.27b: (1) a pin/hinge is a permanent ICR itself, (2) the permanent ICR for a slider is at infinite distance in a direction perpendicular

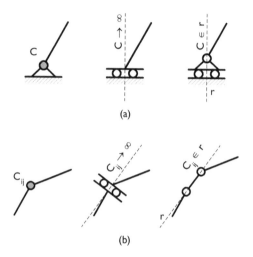

(a)

(b)

Figure 5.27 Instant centre of rotation. (a) Fixed and (b) permanent.

to the sliding motion and (3) the fixed ICR for a pendulum can be any point belonging to its axis.

- **Theorems:** generally, the visual inspection does not allow identifying all the ICRs, and it is possible to locate the remaining ones by means of two theorems: (1) for any two-body structure, the necessary and sufficient conditions for it to be a mechanism is that all the ICRs should be aligned (C_1, C_2 and C_{12}) and (2) for any three-body structure, necessary and sufficient conditions for it to be a mechanism is that all permanent ICRs should be aligned (C_{12}, C_{13} and C_{23}).

  The application of the mentioned theorems is shown in Figure 5.28 regarding an SDOF structure. After the visual inspection, the ICRs C_2 and C_{13} are still unknown, but they are univocally determined thanks to the couples of alignments (i.e. by means of the intersection of two lines). Usually, not all ICRs are required but, for the sake of clarity, C_{13} location is also calculated at infinite distance in the horizontal direction.

- **Components of displacement:** once all the ICRs are determined, it is possible to draw the horizontal and vertical components of the mechanism displacements. Considering the previous structure, the displacement of any part is described according to a single (Lagrangian) parameter, for example the rotation θ of one of the bodies. According to the definition of fixed and permanent ICRs, assigning an arbitrary clockwise rotation to body 1 (in both horizontal and vertical projection), it is possible to draw the entire diagram. In particular, the projections of C_1, C_2 and C_3 have null displacements, whereas C_{12} and C_{13} represent the point of mutual rotation. In this case, the knowledge of C_{13} is superfluous (Figure 5.29).

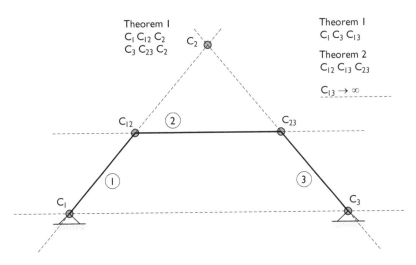

Figure 5.28 Three-body mechanism with explication of the two theorems and the location of all ICRs.

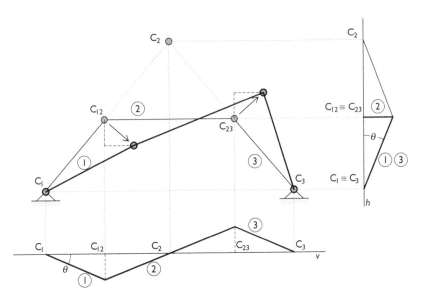

Figure 5.29 Mechanism and its displacement in terms of horizontal and vertical components.

5.2.2.2 Masonry arch

Regarding the study of masonry arches through the kinematic approach, it is possible to proceed as follows. First, the structure should be transformed in an SDOF admissible mechanism through the introduction of a certain number of plastic hinges. This means that, generally, four hinges should be placed in an alternate manner between intrados and extrados. In case of symmetry (of both loads distribution and geometry) five hinges are requested, instead. Note that the releases can also be support displacements instead of hinges, see Como (2013) for examples. Once the hinges are placed, it is possible to locate the ICRs and calculate consequent displacements. Knowing the forces and displacements, the virtual work equation can be written, and the load multiplier associated to the given mechanism can be calculated. According to the kinematic theorem of limit analysis, this value represents an upper bound of the true failure load, which coincides with the minimum of the infinite values resulting from all admissible mechanisms.

Figure 5.30 presents the kinematic analysis of a masonry arch undergoing seismic loads. To create an SDOF mechanism, four hinges are arbitrary placed, as shown in Figure 5.30a. The figure shows also the dead load W applied to the centre of mass of each rigid body, together with the horizontal unknown load proportional to the weight that will induce collapse (λW). Figure 5.30b, instead, shows the location of ICRs and the consequent vertical and horizontal components of displacement. As it is possible to notice, the diagrams are drawn according to a single variable θ that describes the rotation of body 1. Assuming infinitely small rotations, it is possible to assume that the arc of a rotation is equal to the tangent. Therefore, simple calculations can be used to evaluate all other quantities (e.g. the vertical displacement of C_{12} is equal to θ times the horizontal distance between C_1 and C_{12}). Finally, the overall displacement of the mechanism is shown in Figure 5.30c.

Knowing the values and application point of the loads, and their relative displacements, it is possible to compute the virtual work equation (5.1), where L_W and L_S stand for the work done by the weight and seismic load, respectively. Dealing with rigid bodies, the internal virtual work is null.

$$L_W + L_S = 0 \tag{5.1}$$

$$L_W = \sum_{i=1}^{3} W_i v_i \tag{5.2}$$

$$L_S = \lambda \sum_{i=1}^{3} W_i h_i \tag{5.3}$$

where v_i and h_i are the vertical and horizontal centroid displacements computed according to Figure 5.30b. For the given position of plastic hinges (Figure 5.30a), the load multiplier λ of the horizontal loads reads:

$$\lambda = -\frac{\sum W_i v_i}{\sum W_i h_i} \tag{5.4}$$

Since both v_i and h_i depend on θ, it is possible to get rid of this indeterminacy.

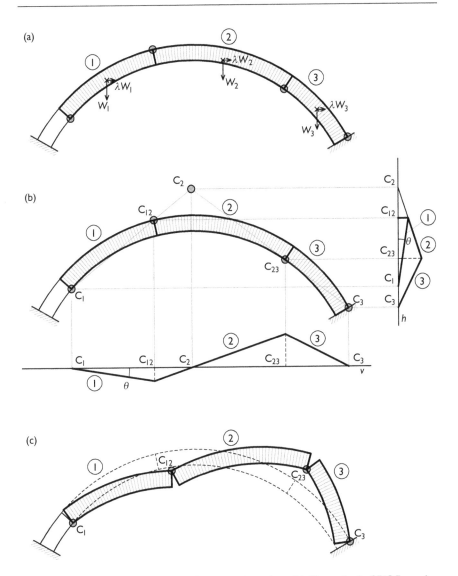

Figure 5.30 Masonry arch subjected to seismic action. (a) Three-body SDOF mechanism and loads, (b) ICRs and displacement components and (c) overall displacement.

A similar analysis can be performed in case a point load λF is applied on the extrados of the arch, but not along the symmetry axis. For the sake of clarity, the same mechanism of the previous case is illustrated in Figure 5.31, where the force and relative displacement v_F are also indicated (the horizontal projection of the displacement is not requested, because there are no horizontal forces applied). According to the assumed location

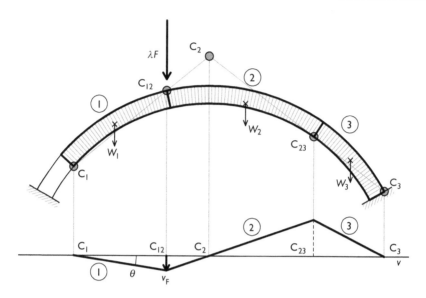

Figure 5.31 Kinematic analysis of a masonry arch with a point load at the extrados.

of the plastic hinges, considering L_F as the work done by the external force, the load multiplier λ can be calculated as follows:

$$L_W + L_F = 0 \tag{5.5}$$

$$L_F = \lambda F v_F \tag{5.6}$$

$$\lambda = -\frac{\sum W_i v_i}{F v_F} \tag{5.7}$$

In case of symmetry of both geometry and load condition, the kinematic analysis can be performed according to Figure 5.32, in perfect analogy with what was already stated (in this case, five hinges are requested to create a mechanism). In particular, it is worth noting that all ICRs and the consequent displacements must be symmetric too; thus, C_2 and C_3 need to be located on a horizontal line passing through C_{23}.

Finally, the kinematic analysis described earlier can be extended to larger structural elements composed by arches. For instance, Figure 5.33 presents the cross section of a masonry arched structure with a point load on the extrados. Possible mechanism types are given in Figure 5.34 together with the relative deformed shapes (of which the third one is frequently the weakest), whose solution is similar to what is described earlier. In particular, it is worth reminding that the true load multiplier (of the point load, in

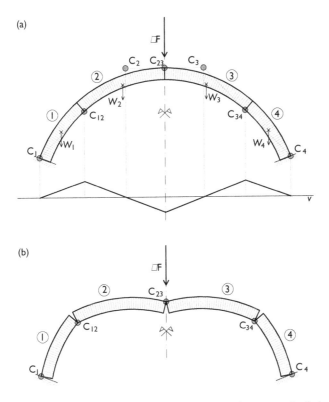

Figure 5.32 Kinematic analysis of a symmetric masonry arch symmetrically loaded.

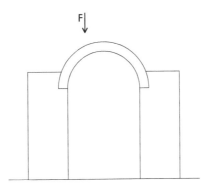

Figure 5.33 Masonry arched structure under eccentric point load.

this case) is the minimum one obtained for all possible locations of plastic hinges in all possible mechanism types.

The same reasoning can be applied if a stiff tie is present (Figure 5.35). In this case, the tie constrains the distance between anchor points **A** and **B** to remain constant. With respect to Figure 5.36, the number of possible

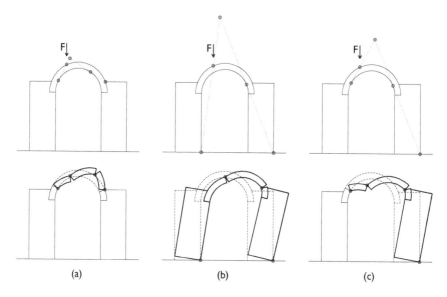

Figure 5.34 Three mechanism types for the solution of the problem in Figure 5.33.
(a) Mechanism type I, (b) mechanism type 2 and (c) mechanism type 3.

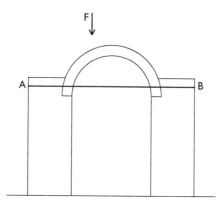

Figure 5.35 Masonry arched structure provided with a tie under eccentric point load.

mechanisms is now restricted. In particular, mechanism type 1 is more likely, type 2 is restricted and type 3 is not possible. The calculations are similar to the previous case, considering that forces in **A** and **B** do not contribute to overall work. The case of a flexible tie would be different as the displacements would not be constrained, and the tie would provide a contribution to the overall work equal to the steel yield force times the tie elongation.

Finally, the cases of two multiple arch systems are presented in Figure 5.37. Under the previous consideration about arched structures, the

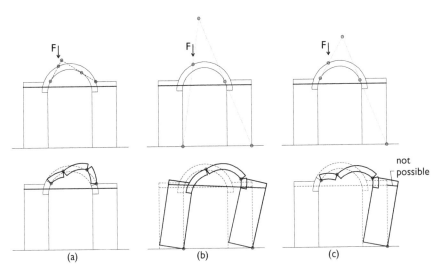

Figure 5.36 Three mechanism types for the solution of the problem in Figure 5.35. (a) Mechanism type 1, (b) mechanism type 2 and (c) mechanism type 3.

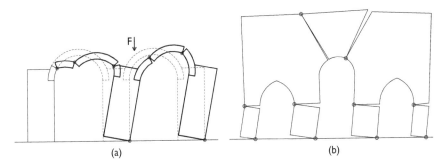

Figure 5.37 Kinematic analysis of two multiple arch systems. (a) Two arches supported by piers and (b) arches with resisting spandrels.

mechanism may involve other parts of the structure. In this case, each individual arch must include at least three blocks and four hinges, whereas each additional arch should add at least two additional moving blocks and three additional hinges (Figure 5.37a). As another example, the failure mechanism of Figure 5.37b shows basically two arched structures connected in the middle by a single pendulum.

5.2.2.3 Masonry macro-elements

Not much different to the static approach, kinematic analysis incorporating the three-dimensional behaviour of vaulted structures is a complicated task.

Still, kinematic analysis reveals itself as an advantageous tool for expedite seismic assessment and strengthening of masonry buildings. In the presence of good-quality masonry with homogeneous constructive characteristic and structural behaviour (that prevent the elements to just crumble away, i.e. material disintegration), when subjected to seismic loading, ancient buildings can be studied as an assembly of independent and considerably autonomous sub-structures called macro-elements. Since these are independent from age, technology, dimensions and overall shape of the building, the relative mechanisms are considered fundamental (National Civil Protection Service, 2013), and they give the possibility to predict the seismic behaviour of a building simply by analogy (Regione Toscana, 2003). Such approach is particularly suitable for traditional masonry buildings, which often do not satisfy the general conditions for the application of standard procedures based on, for example, the 'box' (or integral) behaviour of the structure, rigid diaphragms and floors, in-plane frame-type behaviour of masonry, or, more simply, on modal superposition (or spectral analysis).

The study of macro-elements, only briefly addressed here, is based on the individuation of collapse mechanisms from abacuses relative to different constructive typologies (isolated or aggregate building, churches, etc.).

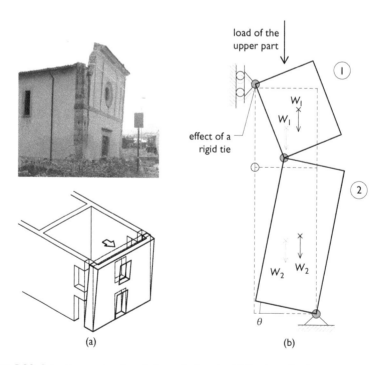

Figure 5.38 Seismic overturning of the main façade. (a) Earthquake damage observation before strengthening and (b) possible failure mechanism after tie strengthening at the top.

Then, the mechanism should be schematized by means of a kinematic model to evaluate the minimum collapse load, given by the load multiplier of the horizontal loads proportional to the weight.

Figure 5.38 provides an example of a mechanism describing the simple overturning of the main façade, easy recognizable from the photos. The kinematic schematization is also proposed for a strengthened system with a tie at the top, in which the new collapse mechanism can be the out-of-plane failure with a hinge line at mid-height of the wall.

Chapter 6

Damage and collapse mechanisms in masonry buildings

Most damage in masonry buildings results from the low tensile strength of the material due to the limited bond between mortar and units. This means that, as soon as tensile stresses arise in structural (and non-structural) elements, cracks normally appear. However, cracking in masonry is an ordinary condition, and it does not necessarily point to structural problems. In fact, separated parts can resist independently (i.e. cracks may be compatible with equilibrium), if an alternative load path can be found, often providing a relatively large capacity. This aspect was already evident in ancient times: Viollet le Duc, for instance, used to (improperly) call this feature 'elasticity' (Di Pasquale, 1996, p. 222). In turn, not all cracked masonry buildings are safe. In addition, other than tension, shear and compression may also induce damage in masonry buildings.

Besides permanent loads induced by gravity, other causes of damages for masonry buildings are possible, such as climatic, anthropogenic, chemical or ageing. The complexity, variety and interaction of these phenomena along with the usual long life of the building are a challenge for those involved in conservation. For this reason, this activity is defined as forensic engineering, that is, the application of engineering principles to the investigation of failures or other performance problems. In fact, the evaluation of the structural safety of historical buildings faces limitations due to the difficulties inherent to an accurate and comprehensive characterization of the causes of damage and decay, as discussed in Section 1.5.

In the present chapter, the main causes of damage are briefly described making use of real cases, referring the reader to literature (Feilden, 2003; Croci, 2008) for further details. Given the fact that most of the existing built heritage is made of masonry and masonry-like materials such as earth, this chapter focuses exclusively on masonry construction.

6.1 LOADS, STRUCTURAL ALTERATIONS AND SETTLEMENTS

6.1.1 Damages in vaulted structures and compatibility cracks

According to the stimulating article of Heyman (1972b), the natural state of the masonry flat arch or lintel is the cracked state. A paradigmatic case of a cracked lintel is shown in Figure 6.1a. On closer inspection, the crack marks the (abrupt) change between two structural resistant forms, that is, from beam to arch (Section 4.2.1). Due to the limited tensile strength of the stone, in fact, the bending capacity of the lintel is very low and, by means of cracking, the structural system switches to a three-hinge arch, which is based on a state of equilibrium involving only compressive stresses. The three hinges are placed at the two internal contact points with the piers (intrados) and at the extrados in correspondence with the crack. The consequent structural scheme guarantees an overall large capacity, providing that the abutments or lateral walls resist to the outward thrust exerted by the embedded arch.

The same reasoning can be applied to the dome of Figure 6.1b. As shown in Section 5.2, due to the rather low tensile strength of masonry, concentric cracks appear transforming the dome in a system of arches embedded into the thickness. The capacity is thus guaranteed by the new system involving only compressive stresses. Note that cracks coincide with the centre of the openings, and the location of ribs demonstrates an evident understanding of force flow by the original builders.

Accordingly, rather than alarming structural problems, cracks often represent a functional expedient of masonry elements to find another resistant scheme. However, if loads increase or change, or the supports undergo sliding, sinking or rotating movements, the cracks might increase in width or quantity, leading to collapse. Looking again at Figure 6.1a, the three hinges

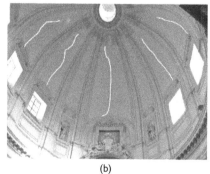

(a)   (b)

Figure 6.1 Cracks in masonry. (a) Lintel and (b) dome.

in the lintel are not sufficient to generate a mechanism, nor the hinges are located in a way the structure may be considered unstable (i.e. alignment of three hinges). In this case, the collapsing mechanism is only partially developed, and the structure is likely to be still safe, at least for gravitational loading. A further hinge, for instance, at the base of one of the piers, may lead to failure.

Figure 6.2 shows two examples of mechanisms involving masonry arches. The first picture shows a real single-span bridge tested up to collapse. As it is known, four hinges are enough to transform the arch into a mechanism, which normally requires large displacements at collapse. In turn, due to the symmetry of the problem (in terms of geometry and load distribution), the mechanism for the arched structure of Figure 6.2b needs the formation of five hinges (e.g. three on the arch and two at the external edge of the abutments). As a general rule, when a mechanism is suspected, a proper historical investigation may shed light on its development: a structure can be safe if it was adequately strengthened by means, for instance, of ties or additional buttresses that 'froze' the displacements, or if soil consolidation further limited progressive movements.

Once a mechanism is activated, a 'ductile' mode of failure generally occurs, involving gradual and visible evidence, namely large displacements. The ductile character of the failure permits, in most cases, to prop the structure and prevent collapse (Figure 6.3).

The mechanisms shown assume that the structure has some integral (or global) behaviour. However, if different parts experience different stress/deformation levels, cracks may appear leading to separation among parts. This type of cracks is often misinterpreted as damage related to collapse mechanism (as in the classical understanding of 'Sabouret' cracks in Gothic cross vaults, see Chapter 5), but, as far as these individual parts can resist independently (i.e. cracks are compatible with equilibrium), the structure can be still considered safe.

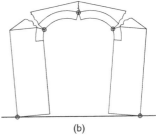

(a)   (b)

Figure 6.2 Collapse mechanisms for masonry arched structures. (a) Single-span bridge and (b) arch and abutments.

Figure 6.3 Activated collapse mechanisms propped with temporary measures.

A clear example of these 'compatibility cracks' can be observed in arches and barrel vaults. Figure 6.4 shows how cracks parallel to the arch development divides the original arch into two main rings, or separate the ring from spandrel wall. These are typical damages in masonry bridges due to compatibility of deformation (e.g. the spandrel wall is far stiffer in its plane than the inner portion of the arch, Figure 6.4b). Cracks may have been originated by settlements and joint alignments, and the capacity of the arch or the spandrel seems not compromised by the crack. Still, both possible failure mechanisms (arch collapse with hinge development and out-of-plane collapse of spandrels) need to be evaluated.

Compatibility cracks may also appear in arch bridges along the abutments, as shown in Figure 6.5a. Taking into account the possible load distribution, the cracks follow the anti-funicular curve drawn in Figure 6.5b;

(a)

(b)

Figure 6.4 Compatibility cracks for masonry arch bridge. (a) The original arch is split into two equally resistant rings and (b) crack between the inner ring and the exterior edge ring (intrados, bottom view).

(a)   (b)

Figure 6.5 Compatibility cracks in the abutment. (a) Actual conditions and (b) schematic thrust line.

the cracks can be due to a shear crack or due to the detachment of the lower part of masonry that does not contribute much to the overall stability of the structure (i.e. it mainly stands independently).

The concept of stress resultant (the thrust line is actually a force resultant) can be applied also to understand the crack pattern in a façade wall. Figure 6.6a shows the typical crack pattern above an opening. This is basically

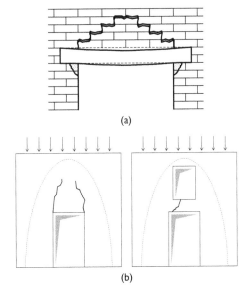

(a)

(b)

Figure 6.6 Compatibility cracks in case of façade wall with openings. (a) Over a lintel and (b) overall view.

due to the flexibility of lintels (often timber ones, but they can also be a cracked stone or masonry) and its natural deflection in the centre, together with the stress distribution over it. The stresses above the opening flows in an arched shape to the lateral piers (see ahead Figure 6.12). Looking at the overall behaviour of the façade, openings may foster the generation of large relieving arches embedded within the fabric (Figure 6.6b). In this regard, according to the geometry of the wall, relieving arches tend to acquire as much rise as possible to minimize the thrust. The unloaded parts, below the relieving arch, may crack, get loose and even might be partially lost, but the main structure may still be safe.

In general, the existing cracks around openings, such as oculi, rose windows and others, give important information about the estimation of the 'true' working arch. Looking at Figure 6.7a, corresponding to a picture of one of the oculi in the basilica of Santa Maria del Mar (Barcelona), the resistant arch is defined by the diagonal cracks in the lower part of the opening and the opening itself. As it is possible to see,

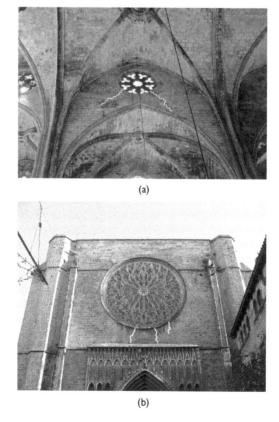

(a)

(b)

Figure 6.7 Compatibility cracks close to oculi and rose windows. (a) Santa Maria del Mar and (b) Santa Maria del Pi, both of them in Barcelona (Spain).

the form of the relieving arch does not coincide with the arch of the clerestory or with the shape of the oculus. Figure 6.7b shows compatibility cracks below the rose window in Santa Maria del Pi in Barcelona (compare with Figure 6.6b, right).

At the scale of the entire building, relieving arches embedded in the structure may still be individuated, together with possible compatibility cracks. Figure 6.8, for instance, shows the case of the Little Hagia Sophia Mosque in Istanbul (before restoration). Analysing the stress distribution, the weight of the main dome is essentially supported by the largest columns that form the high-rise relieving arches (Figure 6.8a). The crack pattern shown in Figure 6.8b is compatible with this interpretation.

6.1.2 Compressive cracking and buckling

Masonry generally shows a good compression strength (still rather variable, from values as low as 1.0 N/mm² and below, for adobe and rubble masonry, to values of 10 N/mm² and higher for good quality ashlar masonry) that guarantees an adequate behaviour under permanent and service loads in the majority of historical buildings. Masonry elements are often oversized in comparison to their compressive strength. Still, note that masonry walls support mostly their self-weight.

When compressive stresses approach about half of the compressive strength, compression cracks tend to appear, as observed in tests. These cracks, parallel to the direction of compressive stresses often occur due to irregularities such as vertical joints, voids, wedges in the horizontal joints or direct contact between stones, natural fracture of stones and foliation or rift direction in stones. These cracks are precursors of failure, occurring at a later stage, thus providing often some early warning. Still, it is worth underlining that compression failure is quasi-brittle, meaning that

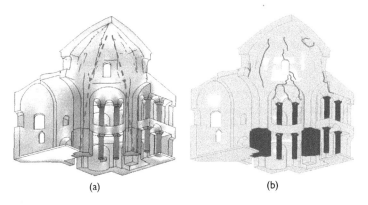

(a)   (b)

Figure 6.8 Compatibility cracks in the Little Hagia Sophia Mosque (before restoration). (a) Schematic relieving arch involving main piers supporting the dome and (b) crack pattern.

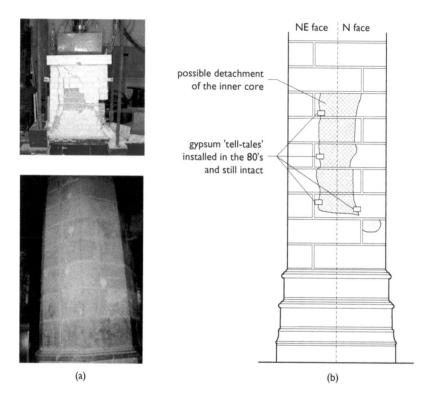

(a)   (b)

Figure 6.9 Compressive cracking on vertical structural elements. (a) Experimental test and pillar in Mallorca Cathedral (b) drawing of the same pillar highlighting its internal damage.

cracks tend to remain with small width or visible deformation up to peak load. After reaching the peak load, rupture is uncontrollable, unless load is removed. As gravity loading will not decrease upon deformation, masonry crushing is likely to compromise the overall stability of the structure (or parts of it) and, in many cases, there will be hardly any possibility to apply preventive emergency measures.

Figure 6.9a shows the results of a compressive test on a small masonry assemblage. At the end of the experiment, vertical cracks are easily noticeable together with partial spalling of corners. By analogy, the same crack pattern can be observed in vertical structural elements due to stress concentrations, for example one of the pillars in Mallorca Cathedral (Figure 6.9b). Similarly, stress concentration can cause also cracks below point loads applied over supports, particularly when applied on top of openings (Figure 6.10).

Similar damage may be found in unconfined parts subjected to high compression stresses, leading to splitting and consequent loss of material. The damage of one corner of a stone capital is shown in Figure 6.11a, and the case of an eccentric rib is presented in Figure 6.11b. This damage can be fostered by

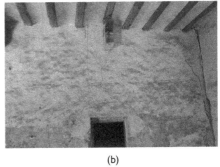

(a)   (b)

Figure 6.10 Two examples of cracking at concentrated load application (helped by an opening below) under (a) roof and (b) floor beams.

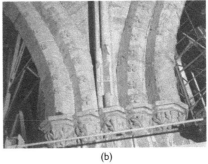

(a)   (b)

Figure 6.11 Compressive cracking for unconfined parts. (a) Cracking and splitting of one corner in a stone capital and (b) loss of material for an eccentric rib.

a combination of several effects, such as natural fracture of the stone, damage induced by stone carving (leading to micro-cracking), thin mortar joints with stress concentration and global movements (e.g. the rotation of a carved ribbed vault leading to compression in the ribs might produce local crushing due to the limited quantity of material available in the eccentric compressed area).

Another example of compressive damage is shown in Figure 6.12 and regards a façade wall provided with openings. The results of the elastic analysis are illustrated in the first picture, where the compressive isostatic lines are drawn. Isostatic lines are lines of constant stress and indicate the trajectories along which pure compression (or tension stresses) flow. One of the main properties of isostatic lines regards the orthogonality between compressive and tensile ones. This means that pure tensile stresses are always perpendicular to compressive isostatic lines and cracks may likely develop along these lines, as shown in Figure 6.12b (similarly to masonry in compression shown in Figure 6.9a), above all when they strongly change curvature.

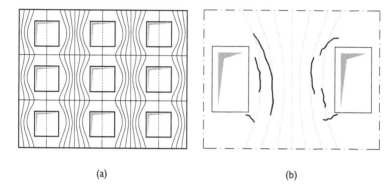

(a)   (b)

Figure 6.12 Compressive cracking due to geometry. (a) Compressive isostatic lines according to elastic analysis, indicating the force flow and (b) real cracks in a pier defined by two windows.

Additionally, compressive loading can lead to the so-called *buckling*, which is inherent to the instability of a slender structural member subjected to compression (the classical example being given by a compressed ruler), in which the compressive stress at failure is lower than the ultimate compressive strength of the material subjected to uniaxial compression. Even if masonry walls are often thick and instability phenomena might seem of limited interest, in many cases it is important to take into account that wall leaves poorly interlocked can suffer instability. Due to transverse tensile stresses, the wall may experience inner cracking and separate (Figure 6.13a shows the example due to the application of an external load). The same can occur in the case of three-leaf walls (Figure 6.13b)

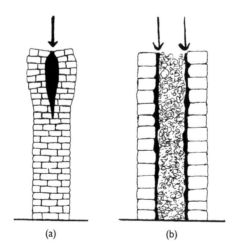

(a)   (b)

Figure 6.13 Inner cracking and leaf separation in a masonry wall. (a) Units poorly interlocked and (b) schematic three-leaf cross section.

with poor connection between leaves, leading to the consequent leaf separation. In general, to prevent instability failure, stones across the full wall section are required to ensure monolithic behaviour or, at least, very good keying of the external leaves in the internal core, with no continuous vertical joint.

Most of the stone-faced medieval and Renaissance walls, in fact, were built with two outer faces of stone ashlar filled with random stone rubble mixed with (frequently) poor mortar, up to 30% of volume. Three leaves are usual for any very thick masonry wall. Many reasons justify separation of leaves, including lime not perfectly carbonated in thick walls, rain water infiltration (may leach out the mortar, may lead to expansion due to freezing–thawing action or may induce water vapour pressure upon dying) and, most important, thermal variations, as the temperature gradient can be very large in thick walls and is a recurrent cyclic phenomenon. Moreover, given the low stiffness of mortar (i.e. relatively soft behaviour), the deformation of the internal fill of a three-leaf wall may transfer the load to the outer sound masonry faces that, in turn, start to work independently, increasing the risk of buckling.

Other examples are shown in Figure 6.14, where the buckling phenomenon is triggered by existing inner cracking or separation.

Buckling can also cause the failure of slender compressed members due to sudden geometric instability, as illustrated in Figure 6.15 (usually in more recent masonry construction and locations where earthquake consideration is not an issue). This phenomenon is considered in all codes for the design of compressed members, made using any material such as concrete, steel, masonry or timber.

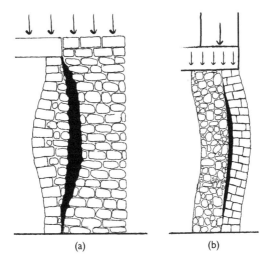

(a)   (b)

Figure 6.14 Buckling triggered by existing inner cracking or separation in case (a) the wall cross section is partially or (b) fully involved in the floor supporting.

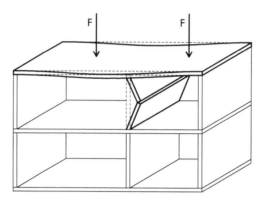

Figure 6.15 Buckling of compressed slender masonry element.

Finally, failure in compression may occur in the long term, even at moderate compressive stresses, due to the decay of the material or long-term damage, usually referred to as creep (Binda, 2008). This type of failure has been observed mostly in heavy towers, as the Campanile of Venice in 1902, the Pavia Civic Tower in 1989, both in Italy, and several other cases in different locations. A recent impressive example is the failure of Noto Cathedral (Italy) in 1996 due to poorly built masonry piers (Figure 6.16). Also, Figure 6.17a shows the partial collapse of the Medieval Maagden tower at Zichem (Belgium) in 2006 attributed to long-term damage accumulation under high stress levels (Verstrynge et al., 2011). The restoration works carried out are illustrated in Figure 6.17b, where the new part is easy identifiable.

6.1.3 Lateral in- and out-of-plane actions

When a building is subjected to lateral loads (e.g. seismic loads or thrust from vaults), according to the arrangement of masonry walls, it is possible to distinguish between in-plane and out-of-plane actions, that is, with applied forces parallel or perpendicular to the middle plane of the wall, respectively.

Regarding in-plane loading, depending also on the compression stresses acting on the wall (due to upper floors, roof or other loads), lateral actions may overcome bond and friction in mortar joints and also lead to toe crushing of masonry or cracking of masonry units, resulting in overall in-plane distortion of the wall and different modes of failure. Figure 6.18 illustrates the main ones, sorted according to increasing vertical load applied on the wall: rocking, sliding (with stepped cracks), diagonal cracking (affecting also the units) and diagonal–compression cracking–crushing.

In turn, out-of-plane actions (such as those due to the horizontal thrust of vaults supported on walls) may contribute to generate out-of-plumb and curvature (in this case, also a possible buckling trigger), as well as associated cracks in walls (Figure 6.19).

(a)

(b)   (c)

Figure 6.16 Collapse of Noto Cathedral (Italy), which occurred in 1996. (a) Overall view; (b) remains of dome and drum; (c) debris following the collapse of central nave and right aisle.

6.1.4 Design, construction and architectural alterations

Inadequate construction practices, neglects and mistakes may cause deformation and damage at either the short or long term. Figure 6.20a shows the effect of incomplete filling of mortar joint that caused cracking in a compressed pier of Mallorca Cathedral. Figure 6.20b presents the results of the early removal of centring (and/or inadequate mortar mixture) in the church of Santa Maria del Mar (Barcelona), which caused mortar expulsion and deformation of the arch. Historical research has shown that the problem was due to the destruction of centring by a fire occurred during

(a)   (b)

Figure 6.17 Maagden tower at Zichem (Belgium). (a) Partial collapse that occurred in 2006 and (b) current aspect after restoration works.

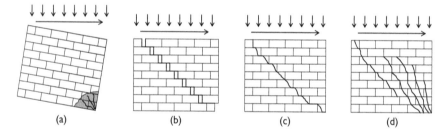

(a)   (b)   (c)   (d)

Figure 6.18 Failure modes due to in-plane action. (a) Rocking, with possible toe crushing at a later stage; (b) frictional sliding (stepped cracks); (c) diagonal cracking involving masonry unit failure; and (d) diagonal–compression cracking–crushing.

the construction. Figure 6.20c shows, instead, the detail of a cross vault (in the right) of Mallorca Cathedral built with no interlocking connection with the transverse arch (in the left). In the long term, the joint between both elements has developed a significant separation. This is a recurrent form of damage due to insufficient interlock between an existing masonry element and a new added element.

Other construction features (such as construction joints) are often incorrectly associated with damage. Construction joints, in fact, mark the boundary between different construction phases. In some cases, they may act as movement joints, sometimes developing into larger cracks affecting other parts of the structure. In this regard, historical and archaeological research may help in the correct interpretation. Figure 6.21 shows two

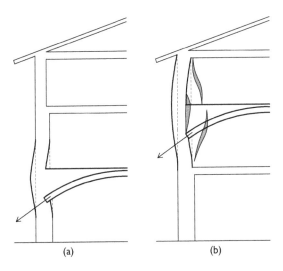

(a)   (b)

Figure 6.19 Out-of-plane action on a perimetral wall. (a) Thrust action and overall displacement and (b) thrust action, overall displacement and additional crack patterns in transverse walls.

examples of construction joints, in Girona Cathedral and in the church of Santa Maria del Mar in Barcelona.

Short- and long-term damages may also be produced by auxiliary devices used during the construction process. Converse to the construction features, which are part of the present structure, construction devices were used as auxiliary tools in the past and may not carry any function at present or may even have been removed. The case of the damaged pier of Mallorca Cathedral is shown in Figure 6.22a. The stone blocks are damaged, among other reasons, due to stress concentrations produced by wooden wedges embedded within the joints. During construction, in fact, wedges were often used to position the stone blocks while the mortar was hardening. Stone and iron wedges are also often found for levelling stone blocks. Another typical example is found in the holes originally used during the construction process to sustain scaffolding and centring. In the church of Santa Maria del Mar (Barcelona), for instance, they were left fully open, passing throughout the wall thickness (Figure 6.22b). Holes may cause stone durability problems due to humidity, vegetation grow, bird nesting, beehives and others (Figure 6.23). Bird dropping, for instance, has an important corrosive effect on monuments and statues.

Along with construction devices, in many cases, the construction process needed to overcome difficult and delicate construction stages in which accidents, large deformation or some damage were possible, and even likely. For instance, the construction process would have required the use of auxiliary members (such as wooden centrings or iron ties). Deformation that occurred in the process would remain frozen in the final geometry of the structure.

Figure 6.20 Inadequate construction practices. (a) Cracks following an incomplete filling of mortar joint, (b) early removal of centring with consequent mortar expulsion and deformation of the arch and (c) separation due to the
absence of interlocking connection in a cross vault and a transverse arch.

In the diagnosis, it is thus essential to identify damage and deformation
developed during the construction process, as they should not be misinterpreted as later damage caused by still active deterioration processes.

Figure 6.24 shows the example of the flying buttresses in Mallorca Cathedral.
According to modern interpretation, their deformation may be largely due to
the deflection of original centring. However, mistaking it for subsequent damage, one of the flying arches was propped during the 18th century.

Finally, later architectural or structural alterations may be a frequent
cause of damage, again in the short or long term. Subsequent transformations and enlargements have been normal practice through construction
history (see Figure 6.25), leading to inadequate alterations such as over-
weight and weak interlocking between different masonries. This is also

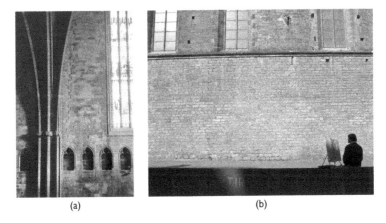

(a)   (b)

Figure 6.21 Construction joints. (a) Girona Cathedral and (b) Santa Maria del Mar (Barcelona).

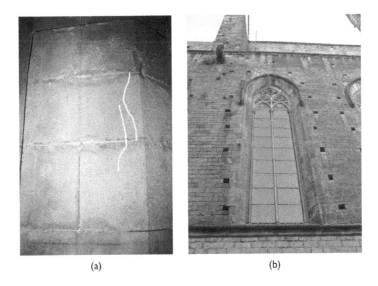

(a)   (b)

Figure 6.22 Damage due to construction devices. (a) Mallorca Cathedral (wooden wedges embedded within the joints) and (b) church of Santa Maria del Mar, Barcelona (unfilled openings for the scaffolding).

the case of the later construction (16th century) of a lantern tower above the medieval structure, which caused severe and persistent structural problems in Tarazona Cathedral (Spain). Figure 6.26 shows two pictures of the building under an intervention to address these problems. Here, the word *cimborio* is used, which means a raised structure like a dome or a cupola, often and more specifically, a lantern built over the crossing of a Gothic cathedral, usually octagonal in plan.

(a)   (b)

Figure 6.23 Biological-induced damage due to open holes in the construction. (a) Vegetation growth and (b) bird nesting.

Figure 6.24 Flying buttress of Mallorca Cathedral propped during the 18th century.

6.1.5 Soil settlements

Problems at the foundation or in the soil may manifest, in some cases, as rigid body motions involving overall out-of-plumbing (rotation), uniform or differential settlements, or lateral displacements. In particular, a rigid rotation of part of the structure may generate overstressing in compression with possible crushing of the material, or overall instability, both of them leading to collapse. As a general rule, the shape of the crack and the

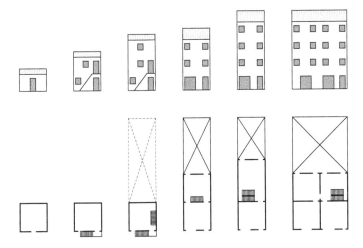

Figure 6.25 Example of a typical architectural and structural alteration throughout centuries for a house in historical centres.

(a)   (b)

Figure 6.26 Tarazona Cathedral (Spain) during the retrofit intervention. (a) External and (b) internal views of the cimborio built during the 16th century above the original medieval structure.

relative movement of the two parts may indicate the type and extension of settlement (Figure 6.27), thus assisting diagnosis and subsequent safety assessment.

Problems related to soil settlement can be especially dangerous for load-bearing wall buildings with low-quality masonry or weak interlocking

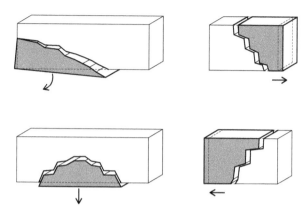

Figure 6.27 Schematic illustrations of cracks and possible causes.

between perpendicular walls. In these cases, settlements may lead to the disintegration of the element or to detachment, rotation and out-of-plumb deformation of single walls. The latter damage may be amplified by out-of-plane thrust of vaults and roofs, with the consequent failure of elements due to excessive deformation (Figure 6.28).

In wall systems of good quality masonry, with the structural elements properly interlocked (as the one schematized in Figure 6.29), differential settlements generate in-plane shear deformation within the walls, resulting in diagonal cracking (usually following mortar joints). Lintels and

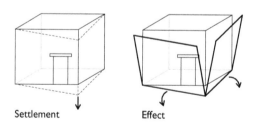

Settlement   Effect

Figure 6.28 Differential settlements for a system of masonry walls with poor interlocking.

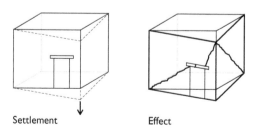

Settlement   Effect

Figure 6.29 Differential settlements for a system of well-connected masonry walls.

spandrels over openings are easily involved in the crack pattern, as they define singular parts and tend to attract damage. In general, the structure may still be stable, even for significant settlements, if no out-of-plumbing of walls is produced. These examples demonstrate the difficulties in forensic engineering as two seemingly equal buildings, subjected to the same phenomenon, can provide a much different response, which also clearly affects the safety level. The damage associated to Figure 6.28 is typically much more dangerous than the damage associated to Figure 6.29.

In case the settlement interests a local area, the structure may crack in different ways according to the thickness and material composition of the element. Regarding the corners, a typical damage is shown in Figure 6.30, where the separation of parts follows the generation of relieving arches. A similar problem can also lead to a different condition of a more extensive damage in the corner area, if masonry bond is more effective (Figure 6.31).

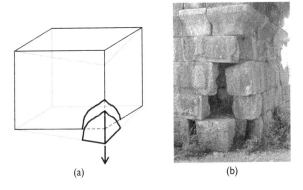

(a)   (b)

Figure 6.30 Vertical settlement at one corner. (a) Schematic view of a relieving arch and (b) observed damage for a masonry building.

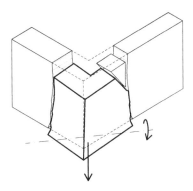

Figure 6.31 Vertical settlement at the corner with overall separation and rotation of the intersecting part of the walls.

Considering the case of vertical settlement at the intersection between orthogonal walls (Figure 6.32), the crack can be easily associated to a relieving arch. On the other hand, a vertical settlement in the middle part of a wall can produce either a relieving arch or a bending failure depending on the geometry involved (Figure 6.33).

Considering the case of an entire building with large openings in the façade and inner load-bearing walls, cracks due to differential settlements tend to concentrate around the openings, usually in the spandrels. Figures 6.34 and 6.35 show the effects of a vertical settlement on one side and in the central part of a building, respectively. Typically, spandrels damaged by soil settlements will experience a cracking pattern, partly or totally developed, including diagonal shear cracks and/or vertical bending cracks.

On the other hand, the outward rotation of one of the façades may induce severe damage to the perpendicular walls. Again, as described in Figure 6.36, the crack pattern is much influenced by the presence of openings.

Differential settlements in buildings may occur due to different pressures induced by the walls and lack of homogeneity in the soil as well as landslides, earthquakes and other geological phenomena. Regarding settlements

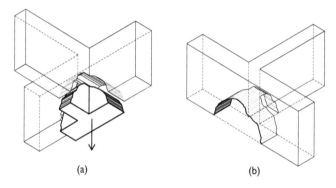

(a)   (b)

Figure 6.32 Vertical settlement at the intersection between orthogonal walls. (a) Bottom view and (b) damage at the intersection.

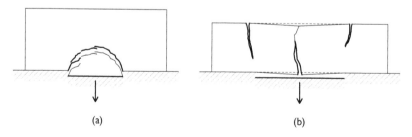

(a)   (b)

Figure 6.33 Vertical settlement in the middle part of a wall. (a) Relieving arch and (b) bending failure.

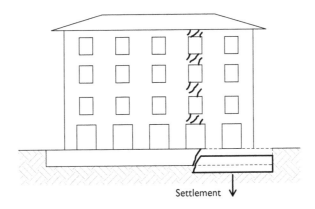

Figure 6.34 Schematic view of a vertical settlement on one side of the building.

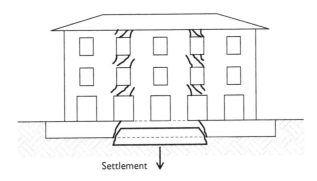

Figure 6.35 Schematic view of a vertical settlement in the central part of the building.

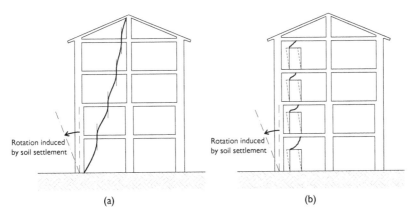

Figure 6.36 Rotation of one of the façades and effects on the perpendicular walls induced by settlement. (a) Solid wall and (b) wall with openings.

in the horizontal direction, in analogy with horizontal loads (see Section 6.1.3), it is possible to distinguish in-plane and out-of-plane settlements. The effects of an in-plane settlement are shown in Figure 6.37 for a solid wall and a wall with openings. Differential horizontal settlements will normally produce single vertical cracks of approximately constant width (as opposite to more complex crack patterns observed for differential vertical settlements, inducing rotations).

In case of out-of-plane horizontal settlements, Figure 6.38 shows possible effects on a solid beam and on a masonry wall with openings, respectively.

Compared to wall systems, arches and arched structures usually show a larger capacity in accommodating large differential settlements. This is due to their ability to develop a sufficient number of hinges that transforms the arch in a static determined system, allowing thus large deformations and delaying the collapse occurrence. Deformations can increase until the ultimate equilibrium condition, which is reached with the generation of the necessary number of hinges (and translational releases introduced by settlements). Therefore,

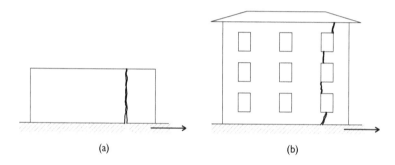

(a)   (b)

Figure 6.37 In-plane horizontal settlement at the end of the wall: effect on (a) solid wall and (b) wall with openings.

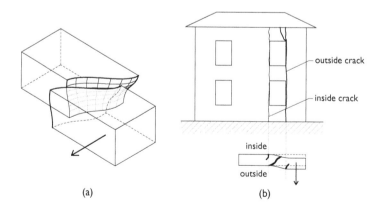

(a)   (b)

Figure 6.38 Schematic view of out-of-plane horizontal settlement for (a) solid beam and (b) masonry wall with openings.

the arch will remain stable, despite the deformation induced by the differential settlement, if it is still possible to fit a thrust line inside the arch.

Next, some examples are given. The effect of supports opening for a masonry arch is shown in Figure 6.39. In particular, the arch accommodates the settlement through the generation of three hinges (four releases, such as hinges or sliding surfaces, determine a mechanism). In case the arch undergoes very large settlement, accommodation may still be possible by developing some sliding at one or more hinges (Figure 6.39b), but no further loading resistance is likely. Figure 6.39c shows a real masonry arch undergoing sliding at supports.

The case of a masonry arch undergoing a vertical moderate settlement at one of the supports is illustrated in Figure 6.40a. Three cases are illustrated: (1) overall rigid body rotation with an internal hinge (normally appearing close to mid-span), (2) two internal hinges and (3) single sliding hinge (likely when rotations are constrained by the rest of the structure). Figure 6.40b shows the nave and the detail of the vaults in the church of Santa Maria del Mar in Barcelona, respectively. In this case, the differential settlement produced a clear sliding at the joint close to the top of the arch.

Similarly, Figure 6.41 presents an example of opening and partial sliding between voussoirs due to differential settlement between supports.

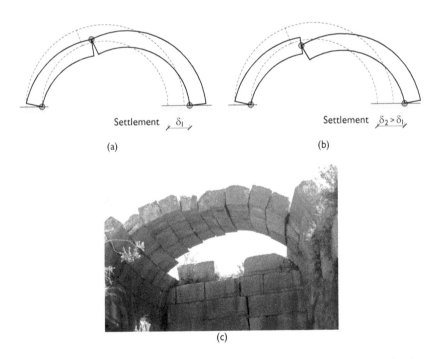

Settlement δ_1

(a)

Settlement $\delta_2 > \delta_1$

(b)

(c)

Figure 6.39 Masonry arch undergoing supports opening. (a) First stage with three hinges, (b) subsequent stage with additional sliding at one or more hinges (four releases are needed for collapse, not necessarily hinges) and (c) real case in the archaeological remains of Myra, in Turkey.

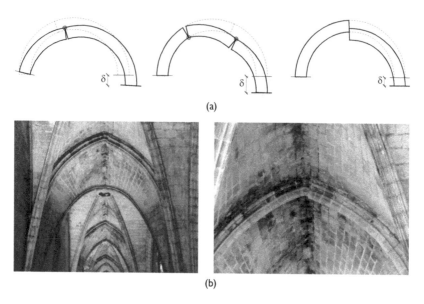

(a)

(b)

Figure 6.40 Masonry arch undergoing vertical settlement at one of the supports. (a) Accommodation through one small internal rotation, development of two internal hinges or through sliding and (b) nave and detail of the aisle vaults in the church of Santa Maria del Mar in Barcelona with pronounced sliding at the top of one of the arches.

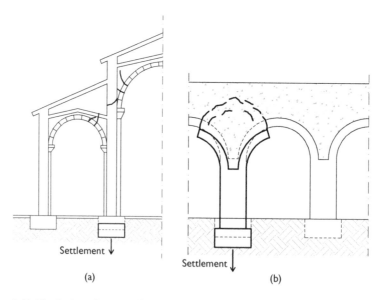

(a)   (b)

Figure 6.41 Vertical settlement of one pier. (a) Cross section of a church and (b) series of arches.

Figure 6.41a shows a transverse cross section of a church. In Figure 6.41b, a relieving arch appears in the material located over a settling pier in a series of consecutive arches.

Severe differential settlements or settlements with a large extension may result in the development of cracks that separate large parts of the construction, the so-called *fragmentation*. This kind of cracks allows independent settlements and, as far as the parts are individually stable, the structure is likely to be safe. This expedient prevents more damages to occur in other parts of the building, as done today with movement joints in modern construction. As already stressed, also fragmentation cracks take advantage of openings and weak parts.

A typical fragmentation case regards the separation of the tower from the rest of the church due to either overall rotation or large settlement (being taller and heavier, the soil foundation of the tower, but also the masonry, is more stressed). Figures 6.42 and 6.43 present two clear examples of fragmentation cracks involving the rotation of towers. As it is visible, the cracks pass through the openings of the façade. Note that, in the design of a modern building, as stated, it is usual to place movement joints between a tower and a low-rise building to prevent this damage.

(a) (b)

Figure 6.42 Fragmentation cracks in the church of Sant Pere Apòstol in Cobera d'Ebre (Spain) involving the tower separation. (a) External view and (b) detail of the façade.

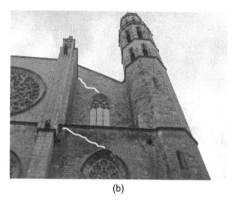

(a)   (b)

Figure 6.43 Fragmentation cracks in the church of Santa Maria del Mar in Barcelona involving the tower separation. (a) Internal and (b) external views.

6.1.6 Large deformations

Due to a single cause or a combination of causes of damage, as addressed earlier, ancient structures may show large, even dramatic, deformations several orders of magnitude larger than those that can be predicted by a standard structural analysis. Being a progressive process, the more ancient the structure, the larger is often the deformation, although it might be no longer progressing. Figure 6.44 shows two examples from Mallorca Cathedral (Spain) and Hagia Sophia (Turkey). Depending on their magnitude, large deformations may need to be taken into account in structural analysis.

(a)   (b)

Figure 6.44 Large deformations in historical buildings. (a) One of the piers in Mallorca Cathedral (Spain) and (b) deformation of arches and columns in Hagia Sophia (Turkey).

6.2 ENVIRONMENT AND ANTHROPOGENIC ACTIONS

6.2.1 Cyclic actions and climatic change

Daily and annual cycles of temperature and humidity (as well as water table level) may produce repeated alterations at the level of materials and structure. This type of alterations is mostly reversible (i.e. the initial condition is recovered after each cycle), but an irreversible component is likely to occur after each cycle. Accordingly, the repetition of multiple cycles along the life of the structure may produce a meaningful cumulative result, contributing, in turn, to additional deformations or damage. For instance, temperature changes generate cyclic expansion and contraction deformations, thus possible cracks and movements that, in turn, permit water infiltration. On the other hand, individual and large cracks may behave as expansion joints for large masonry buildings, providing a relieving joint that actually works as a true feature for the building.

One of the most typical damages due to climate action is related to freezing–thawing cycles of the water penetrated into the structure through pores, cracks and joints. When the water freezes it expands generating cracks, loss of mortar in joints, and even spalling and disintegration of units (Figure 6.45). The accumulation of freezing–thawing cycles may have severe eroding effects on ancient masonry structures. Figure 6.46, for instance, shows the damages of the arch of the Roman Bath in Caldes de Malavella (Spain): although very ancient and robust, the structure has experienced severe deterioration, even in a temperate climate, due to the long-term accumulated effect of thermal and freezing–thawing cycles. These effects are likely to appear together with chemical and biological agents, namely efflorescence (see Section 6.2.4).

(a) (b)

Figure 6.45 Effects of freezing–thawing cycles on brick walls. (a) Spalling and disintegration of units and (b) cracking.

(a)   (b)

Figure 6.46 Effects of hygrothermal effects and freezing–thawing cycles for the Roman Bath in Caldes de Malavella (Spain). (a) Overall view of the arch and (b) cracks in the voussoirs.

Another cause of deterioration for historical buildings is climate change, which has speeded up at an unexpected rate in the last decades and is receiving much attention. Extreme events such as hurricanes, storm surges, heavy downpours, floods and draughts have increased in the recent decades. Vernacular structures, due to their intrinsic fragility, are severely threatened by global environmental changes and they are usually the first type of structures to be destroyed when major natural disasters occur (Figure 6.47a). Likewise, the polar heritage is experiencing several problems due to climate change (Figure 6.47b). Structures of the 'heroic era' of explorers of Antarctica and the huts of trappers in Arctic areas, the remains of factories, machinery and equipment from the whaling and sealing eras as well as historic crosses, graves and other memorials, which mark the sites of historic events, may be lost due to the rising of temperatures. Coastal built heritage is at significant risk due to erosion and sea level rise (Figure 6.47c), including not only piers and lighthouses but also many buildings and transportation infrastructure on shore. According to United Nations Organization, about 10% of the world population live in coastal areas that are less than 10 m above sea level (i.e. 10% of the built environment). Another type of construction that may experience severe problems due to climatic change are earthen structures (like adobe or rammed-earth ones), as their adequate preservation requires the maintenance of appropriate environmental humidity and temperature conditions (Figure 6.47d).

6.2.2 Physical weathering and natural disasters

Water erosion may be considered as the main cause of physical weathering. The persistent friction effect caused by running water, combined with possible dissolution or alteration of the constituents of the material by chemical reactions, has led to large erosion, causing progressive

Figure 6.47 Effects of climate change. (a) Old homes in the Egyptian desert town of Siwa oasis; (b) polar heritage Scott's hut built in 1911 for the British Antarctic Expedition of 1910–1913; (c) 2008 flood in Venice; and (d) earthen watchtower in Dunhuang, China.

deterioration and loss of material. In the occasion of floods, for instance, the effect of friction and impact of sand, rubble, salts and vegetation dragged by water, can contribute to the erosion of the lower parts of structural elements, such as piers of masonry bridges. Further deterioration and undermining at foundations (scouring) may cause settlement with cracks and even collapse. On the other hand, rain contributes to supply humidity to the exposed parts of the structures, fostering damage due to freezing, chemical attack and biological growth. Parts in contact with drain water experience the same effects of running water.

Figure 6.48a shows the erosion and loss of material at the foundation of one of the piers in the Pont Nou in Manresa, Spain. Consequently, the pier exhibits damage, as presented in Figure 6.48b. Starting from the bottom, it is possible to individuate the loss of mortar and stone deterioration due to running water and humidity. Cracks caused by settlements at the base

(a)   (b)

Figure 6.48 Stratified effects connected to water erosion in a pier of Pont Nou in
Manresa (Spain). (a) Erosion and loss of material at the foundation and
(b) stratified deterioration effects.

are evident in the middle part of the pier. Finally, the top of the pier suffers
damage due to biological growth, salt crystallization (efflorescence) and
other effects fostered by humidity provided by rain water that infiltrates
through the deck.

Figure 6.49 presents an example of a partial bridge collapse due to scour.
As discussed earlier, a masonry arch can accommodate settlements at the
supports without failure, thanks to its structural redundancy, as it becomes

Figure 6.49 Examples of a bridge collapse due to scour in North Yorkshire in the UK,
with the loss of the missing central pier.

structurally determined. The failure presented is an extreme case, due to the loss of one support.

In a process similar to flood erosion, for many years, wind erosion has been considered an important cause of decay above all in case of ancient monuments in sandy places (e.g. Egypt). Nowadays, its effect is considered to be of limited importance, certainly minor compared with that of floods, thermal cycles in the long term, earthquakes and others.

One example of damage attributed to wind erosion was the Sphinx in Egypt (Figure 6.50a). In reality, based upon geological considerations, the extreme erosion on the body of the Sphinx might not be the result of wind and sand, but of past floods and salt crystallisation induced by humidity and temperature fluctuations. This may be confirmed by the observation that wind erosion cannot take place when the body of the Sphinx is covered by sand, and the Sphinx has been in this condition for almost the last 4,500 years. Figure 6.50b shows a picture of the condition of the Sphinx in the late 19th century. Furthermore, such erosion has not been found on other Egyptian monuments built of similar materials and exposed to wind for a similar period of time. In this regard, it is widely assumed that wind and sand have done little more than scour clean the surface of the exterior stones.

As a natural hazard, lightning is probably the most frequent cause of violent damage, producing alterations and destruction at the top of tall elements such as pinnacles, domes, spires and towers. Damage is normally local and results from a combination of kinematic action (strike, representing a dynamic mechanical action) and temperature (strong heat leading to spalling and cracking) effects. Severe lightning strikes are said to have destroyed masonry towers. In this regard, Figure 6.51a shows the possible effects of a lightning on the tower of Mallorca Cathedral causing the displacement of one of the blocks. Figure 6.51b,c shows two buildings recently hit in Portugal where important damage was found. In particular, the south tower of Porto Cathedral was struck by lightning at least three times during its history, according to historical sources, always with considerable

(a)   (b)

Figure 6.50 Erosion of the Sphinx in Egypt due to groundwater rather than wind. (a) Picture of the frontal part and (b) picture of the condition in the late 19th century.

Figure 6.51 Lightning effects. (a) Possible effects at the top of a tower of the façade in Mallorca Cathedral (Spain) and detail of a displaced stone block, (b) masonry chimney after the hit and large void at its bottom (2012, Guimarães, Portugal) and (c) Porto Cathedral (Portugal) with the loss of the pinnacle and damage in flying buttress (1951, photo originally torn).

damage. Thus, the installation of lightning protection systems in existing tall building is a demand in conservation works.

Finally, natural disasters (now in some cases fostered by climate change), including floods, hurricanes and tsunamis, represent the most impactful cause of damage, having caused significant destruction of monuments during the past centuries. Natural disasters require emergency actions on a case-by-case basis in order to save and restore damaged monuments or to stabilize the remains of those partially destroyed. Ideally, a suitable risk management plan should exist, and mitigation should be implemented when risk level is not acceptable. In the next subsection, the effect of earthquakes in historical buildings is briefly revised, as it is a major source of loss for the built heritage.

6.2.3 Earthquakes

An earthquake is the shaking of part of the Earth's surface that may induce large amplitude oscillations in the buildings, thus causing significant damage or even destruction of the entire construction. Emergency actions are usually needed for the conservation of built cultural heritage after strong earthquakes.

Earthquake protection plays a key role in the safeguard of historical and traditional structures, whose behaviour is influenced by many features. For any given building, in fact, local seismicity, construction features, local construction technology, alterations experienced along its history, existing damage and condition, geometrical or constructional relationship with other buildings are a few aspects that the analyst should take into consideration. For this reason, structural analyses are usually employed for safety assessment, whereas seismic protection and seismic strengthening measures represent important fields of research.

Some examples of the damage provoked by earthquakes are illustrated next. Figure 6.52 shows the effect of the Lisbon earthquake in 1755, which destroyed a large part of the Portuguese city, with social

(a)   (b)

Figure 6.52 Remains of the Carmo Convent extremely damaged by the Lisbon earthquake in 1755. (a) Stabilized front door with an internal arch and (b) current condition of the main nave.

and philosophical implications all around Europe. To many, this event is considered the starting point of modern seismic engineering. Carmo Convent is shown, which was destroyed by the earthquake and preserved since then in the form of remains to remind the population of the extreme consequences of an earthquake. On the other hand, Figure 6.53 presents examples of Italian churches extensively damaged by the L'Aquila earthquake in 2009. Finally, the effect of the earthquake that struck Bam citadel (Iran) in 2003 is shown in Figure 6.54. This is a recent example of UNESCO (United Nations Educational, Scientific and Cultural Organization) world heritage site raised to the ground in a blink of an eye.

As already stressed in Chapter 5, the observation of seismic damage demonstrated that masonry historic buildings collapse with independent and considerably autonomous mechanisms. Starting from the work of Giuffrè (1993), several studies have investigated this approach in depth. Figure 6.55 shows the out-of-plane collapse mechanisms of existing masonry buildings according to the Italian code of practice (Regione Marche, CNR and University of L'Aquila, 2007). The mechanisms clearly demonstrate the relevance of the position of openings and that of the house within the urban texture in the failure.

(a)

(b)   (c)

Figure 6.53 Examples of damaged churches following the L'Aquila earthquake in 2009 (Italy). (a) Extensive diagonal cracks around the main door; (b) and (c) overturning of the upper part of the façade due to pounding of the ridge beam.

(a)

(b)

Figure 6.54 Bam citadel (Iran). (a) Before and (b) after the earthquake that occurred in 2003.

6.2.4 Chemical and biological agents

Chemical attack results from any substance that reacts with the basic material of a structure. Important sources of attack are microorganisms and a variety of other biological agents such as bacteria, lichens, algae, moulds and fungi, mosses, and liverworts, which may cause deterioration of the surface, deep decay and even disintegration or loss of material. As far as masonry is concerned, decay can affect the unit (stone or brick), mortar joints or both. In turn, wood may also be attacked by other biological agents, including fungi, beetles and termites. The likelihood of biological attack in wood is very much dependent on the humidity level. A wood moisture content below 20% is likely to guarantee protection for fungal attack, while a moisture content below 12% means that eggs and young larvae of beetles will not survive. Termites are more complex, and there are four major types of termites: subterranean, formosan, dampwood and

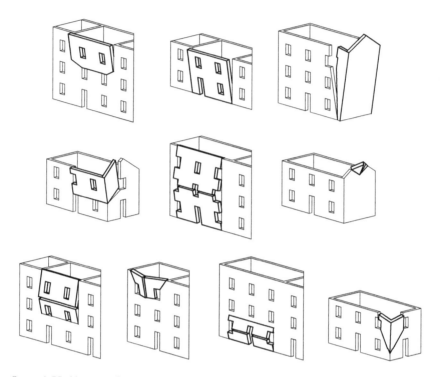

Figure 6.55 Abacus of the out-of-plane collapse mechanisms of existing masonry buildings after (Regione Marche, CNR and University of L'Aquila, 2007).

drywood. Total immersion in water (e.g. in foundations) can also ensure material durability, as oxygen is not available. Still, wood can undergo slow bacterial degradation in fresh water or be attacked by marine borers in brackish or salt water.

In general, the presence of water, in any of its form, causes or accelerates decay of most building materials due to chemical contents (e.g. sulphuric acid, calcium sulphate and dissolved salts) and crystallization process. Besides the direct contact with rain or water disposal, ingress in the building elements can occur through the phenomenon of capillarity, which is the ability of water to rise through a porous material due to surface tension. Capillarity mainly depends on the porosity of mortar and units (if the pores are too large, e.g. gravel or coarse sand, or there are no pores, e.g. steel or glass, then capillarity is not possible; if the pore size is small, such as concrete, mortar, bricks and most stones, capillarity is possible). The height of the capillarity rise is influenced by factors such as pore size (i.e. smaller pores will raise the zone of dampness), the rate of evaporation (i.e. a high temperature and low relative humidity will lower the zone of dampness) and the water table level (i.e. a wet season will raise the zone of dampness), being not uncommon to reach a height of 1.0 m and higher (certainly much more in case of a retaining wall or buried space). Finally, other ways for water to gain access to masonry materials are either by condensation or deposition of aerosol (such as mist, fog or salt spray), which are quite typical of zones close to the seaside.

A chemical and biological attack manifests itself with an alteration of appearance of the material surface (with respect to the original one). The typical cases include *discoloration* (fading, moist spots and staining), *deposit* (soiling, encrustation, efflorescence and biological growth) and *transformation* (patina and crust). Disregarding the self-explaining discoloration, deposit concerns the accumulation of endogenous or exogenous material on the surface without any chemical transformation (Figure 6.56a). On the other hand, patina adheres well to the elements, thanks to chemical reactions (Figure 6.56b).

A more pronounced decay that involves the loss of integrity for the building materials is usually addressed as disintegration. This type of damage can occur inside the material (e.g. layering and loss of cohesion) or between two adjoined materials (e.g. detachment and loss of adhesion). The main cases of disintegration are synthesized as follows: *layering* (delamination, exfoliation, spalling and scaling), *detachment* (loss of adhesion, blistering and peeling) and *loss of cohesion* (chalking, powdering, crumbling, sanding, bursting, alveolization, etc.).

Figure 6.57 shows two examples of a crusted stone capital. This is revealed by the generation of a chemically transformed surface layer not well adhered to the underlying element, which tends to separate due to thermal variations and/or water vapour pressure. The stones of Segovia Aqueduct in Spain are affected by a general loss of cohesion and progressive loss of surface material

(a)   (b)

Figure 6.56 Altered appearance of masonry surface due to (a) deposit and (b) patina.

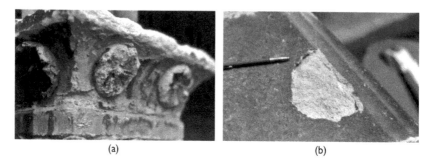

Figure 6.57 Crusted stone capital. (a) Decoration detail and (b) detachment of the external layer.

(Figure 6.58a). As a consequence, the surface of the stone is round and irregular, which can cause stress concentration and cracking. This process might have been fostered, both chemically and mechanically, by modern traffic vibration (usually not an issue, as quite high vibrations are needed to produce damage) and pollution (usually, this can be a severe problem due to car exhaust gases or industry, now often less relevant in developed countries due to the strict environmental requirements). Finally, Figure 6.58b shows how stones subjected to similar conditions, but having (sometimes slightly) different mechanical features, may show a different response in the long term. Note also the evident coexistence of ancient and new stone blocks, which occurs recurrently due to ancient repairs.

Another example of loss of cohesion is shown in Figure 6.59, for the Tarazona Cathedral in Spain. The picture (taken during conservation works) presents clear deterioration and loss of material at the base of a pier.

Figure 6.58 Examples of loss of cohesion. (a) Stones of Segovia Aqueduct (Spain) and (b) response of slightly different stones to similar conditions, including a recent replacement.

Figure 6.59 Severe deterioration due to loss of cohesion at the base of a pier in Tarazona Cathedral (Spain).

This was probably due to the loss of cohesion caused by capillary humidity. Eccentric compression may have acted in synergy to accelerate the process in the more stressed region.

Figure 6.60 shows two examples of alveolization. The first one regards a stone masonry column with a honeycomb decay pattern, probably affecting different components of the material. The second picture shows a Gothic tracery deteriorated by a combination of alveolization and other forms of chemical attack.

A different problem, taking advantage of humidity trapped in masonry (in deteriorated stone, rubble and joints) and accumulation of dirt in cavities due to cracks or voids, is that plants may easily grow in abandoned structures, or any building lacking maintenance. Growing vegetation mostly deteriorates, cracks and dismantles masonry by root and branch penetration and growth, chemical attack due to permanent humidity and organic material and other related effects. In some cases, roots tie together a deteriorated and loosened material with a positive effect, delaying collapse. This is the case of the familiar ivy growing on ancient constructions or large trees growing within masonry, whose removal may affect the stability. In general, although it may be pleasant and romantic (much aligned with the anti-restoration movement of Section 2.2.3), vegetation in masonry indicates abandonment or poor maintenance.

(a)   (b)

Figure 6.60 Examples of alveolization. (a) Stone masonry column severely damaged and
(b) Gothic tracery damaged also by chemical attack.

Figure 6.61 presents examples, before the intervention works, of vegetation in an abandoned convent that was slowly taken by vegetation as the roofs and plasters were damaged and partly lost, and in a medieval bridge in use at the time the photo was taken. Old arch bridges, often with unfilled joints or cracks, having lost their importance in the road network, handed over to local authorities and even decommissioned, face huge maintenance challenges due to the lack of financial and technical resources.

The comparison of the pictures, pre- and post-cleaning intervention presented in Figure 6.62, underlines the former presence of vegetation in the Pont Nou bridge in Manresa (Spain). Diversely, Figure 6.63 shows two impressive examples of robust vegetation within buildings. In particular, the first picture shows the vegetation grown in the temple complex of Angkor Vat in Cambodia. The trees are part of temple, and their removal is hardly possible or extremely risky. According to the actual conditions, in fact, the vegetation exerts a tying effect in the structure, and its loss may lead to the collapse of construction, requiring a stabilizing system or rebuilding. The second picture is similar and shows the Palenque ruins in Mexico.

6.2.5 Fire

Fire resistance of stone, brick and mortar, considered individually, depends mostly on their porosity, being generally mortar the weakest component. Fire can cause mechanical stress and chemical changes inside any of the

Figure 6.61 Vegetation growth in Portugal. (a) Saint Francis Convent in Braga (16th–18th centuries) and (b) Ponte de Vilela.

masonry components, which may decrease mechanical properties such as strength or modulus of elasticity. But masonry heats up from the external areas, causing drainage and evaporation processes to take place. This causes steam pressure, leading to explosive cracking of masonry, known as spalling, which is the major damage manifestation. In addition, when subjected to high temperature, given the different thermal and mechanical properties, units and mortar may tend to behave in a different way, with possibly loss of mortar and cracking. Cracking may also be fostered by constrained expansion of blocks within the structure, especially in compressed members.

(a)   (b)

Figure 6.62 Vegetation growth on the Pont Nou bridge in Manresa (Spain). (a) Pre- and (b) post-cleaning intervention.

(a)

(b)

Figure 6.63 Robust vegetation within the structure. (a) Temple complex of Angkor Vat (Cambodia) and (b) Palenque ruins (Mexico).

The effect of fire in the keystone of one vault in Santa Maria del Mar church (Barcelona) is illustrated in Figure 6.64a. In this case, the flames and the high temperature changed the colour of the masonry surface, with some crusts and loss of material. Figure 6.64b, instead, shows the loss of mortar and stone spalling in a pillar of the same church. The losses were plastered with cement mortar a few decades ago. The intervention was not efficient due to insufficient bond and subsequent detachment of the mortar utilized.

Figure 6.65 shows the effects of fire in the São Domingos church in Lisbon. Two firemen were killed with the collapse of the choir roof. The fire was already strong in the interior when it was noticed. Once the doors were opened, the draft and chimney effect turned it to a tragic accident, with the windows blown away and collapse of the main timber roof. All the interior was lost, and the condition of the church today allows witnessing the impressive result on the stone surface.

In many historical buildings with load-bearing walls, floor slabs consisting of timber or steel joists play a stabilizing role (in the out-of-plane direction) for the walls. Accordingly, the loss of timber or steel beams in a fire may lead to instability of a single wall that triggers a domino effect until the failure of the entire construction. The destruction by fire of roof structures may cause a similar result, and their collapse may drag down all the floor slabs below. Moreover, it is noted that these collapses are sudden, and, as a consequence, they induce important dynamic loads on the walls, often tearing them down or leading to partial collapse due to

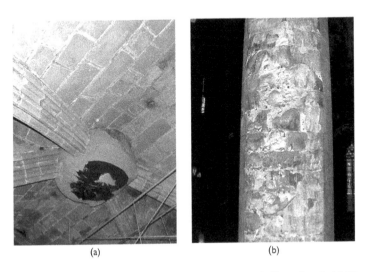

(a)   (b)

Figure 6.64 Effects of fire on the Santa Maria del Mar church (Barcelona). (a) Change of colour and loss of material in the keystone of one vault and (b) pillar with loss of mortar and stone spalling.

(a)

(b)

Figure 6.65 Effects of fire of 1959 on São Domingos Church (Lisbon). (a) General view of nave and (b) extensive stone spalling.

the failure of connections between walls and floors. Examples of masonry buildings heavily damaged by fire are shown in Figure 6.66. As it is possible to see, the fire has burnt and made the inner part of the building collapse, leaving the external walls with no perpendicular supports and prone to instability.

(a)   (b)

Figure 6.66 Effects of fire on masonry buildings. (a) External wall with no perpendicular supports, Great Chicago fire (1871) and (b) collapse of the main façade due to lateral instability, Rochester, NY fire (1904).

The next three figures illustrate that fire remains a major challenge and has a destructive power in the 21st century, with recurring losses across the world. The fires on Quebec City Armoury (2008) and National Museum in Rio de Janeiro (2018) are shown in Figures 6.67 and 6.68, respectively. For the latter, besides the building completely destroyed, more than 90% of its archive of 20 million items have been lost. Figure 6.69 shows the case of Notre Dame de Paris (2019). The blaze destroyed the spire designed by Viollet-le-Duc and much of the timber roof. In addition, works of art, musical instruments, religious relics and woodworks may have been damaged either by the heat or the enormous amount of water needed to fight the fire.

6.2.6 Anthropogenic actions

The last cause for damage in historical buildings is given by anthropogenic actions collected as follows, and illustrated next by a set of examples:

- (inadequate) historical alterations
- lack of maintenance and conservation
- inadequate restorations
- inadequate reconstruction
- destruction: wars and conflicts
- modern living and urbanization (e.g. vibrations and pollution due to heavy traffic)
- tourism and cultural attraction.

(a)

(b)

Figure 6.67 Fire on Quebec City Armoury, 2008. (a) During (© Bernard Bastien/SPIQ.ca) and (b) after fire, with almost full loss of the interior.

Regarding inadequate historical alterations, together with the already-described cimborio (Figure 6.26), also the piers of Tarazona Cathedral (Spain) underwent critical interventions. To make space for a timber choir, the cross section of the piers was strongly reduced during the 16th century, causing deformation and damage to the structure. Figure 6.70a shows the columns and impressive propping visible before the restoration. Another example is presented in Figure 6.70b, with a crack produced by an iron insertion in a pier of Mallorca Cathedral in Spain.

Tarazona Cathedral is also the protagonist of two other inadequate interventions. Figure 6.71a shows the inadequate stabilization (implemented in the 1960s) of one of the flying arches. As it totally lies on the props, it does not generate any lateral force (not even the minimal one) for balancing the thrust of the main nave vaults. The structure actually deformed and cracked due to it. In turn, the full dismantling and reconstruction of the flying arches of Figure 6.71b represent a case of inadequate restoration. During the intervention, in fact, the structure was

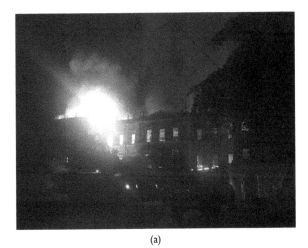

(a)

(b)

Figure 6.68 Fire on the National Museum of Rio de Janeiro on September 2018. (a) During and (b) the day after the fire.

temporarily left without flying arches, and no auxiliary operations were carried out to balance the lateral thrust of the high vaults. Due to this inadequate intervention, the structure experienced additional cracking and deformation (Roca, 2001).

Along the centuries, monuments have constituted a likely target in case of war or religious/cultural conflict. This is certainly due to their importance for people as landmarks of religion, culture and identity. Figure 6.72 shows the cases of the Bamian Buddhas in Afghanistan (destroyed in 2001) and the Old Mostar Bridge in Bosnia and Herzegovina (destroyed in 1993, later reconstructed). The case of the Temple of Bel at Palmyra in Syria (2015) is also shown. Figure 6.73, instead, shows the damaged walls of the church of Santa Maria del Mar (Barcelona) caused by shell fragments and shrapnel during the siege of Barcelona (1713–1714) at the end of the War of the Spanish Succession.

(a)   *(b)*

Figure 6.69 Fire on Notre Dame de Paris on April 2019. (a) During the fire and (b) comparison between before and after.

(a)   *(b)*

Figure 6.70 Inadequate historical alterations. (a) Thinned pier (16th century) in Tarazona Cathedral (Spain) and (b) crack due to the iron insertion in the compressed pier of Mallorca Cathedral (Spain).

Modern living and urbanization may represent an additional cause of damage for monuments and historical constructions. Urbanization, in fact, may threat the original context and even the monument itself. For instance, rural vernacular architecture is often rejected as obsolete (compared with modern standards) rather than considered as a heritage worth to

(a)   (b)

Figure 6.71 Inadequate stabilization and restoration in Tarazona Cathedral (Spain). (a) A flying arch resting on props and (b) full dismantling and reconstruction of flying arches and buttresses.

be preserved. Figure 6.74 presents two examples of urban alteration of the monument context. Figure 6.74a shows the works undertaken for the construction of the metro line in Rome during 1940–1950s in front of the Arch of Constantine and close to the Colosseum. Besides the possible effects of excavations (together with pile driving and passage of heavy vehicles), the daily underground passage of trains produces vibrations in the monuments (note that this becomes less and less relevant for modern metro lines, as the rail tracks are less irregular, the train compositions are lighter and isolation pads are often used, meaning that vibrations tend to be very low). Much more relevant issues might be settlements during station excavation, construction of access points and loss of archaeological buried remains.

Modern urbanization may also cause the loss of the original context and the addition of surrounding buildings not being at the scale of the ancient one. Figure 6.74b shows the case of a masonry church completely surrounded by a modern context of skyscrapers (note also the cars parked in front of the main entrance).

Finally, mass tourism currently threats the conditions of many monuments, historical city centres, historic small villages and archaeological sites due to the deterioration and transformation that it conveys and possible gentrification processes. Accordingly, the possibility of severely limiting the number of visitors is seriously considered in many important

Figure 6.72 Destruction of monuments due to war or religious/cultural conflict. (a) Before and (b) after destruction of Bamian Buddhas in 2001 (Afghanistan); (c) before destruction, (d) during reconstruction and (e) modern aspect of the Old Mostar Bridge in 1993 (Bosnia and Herzegovina).

(Continued)

(f)

(g)

Figure 6.72 (CONTINUED) Destruction of monuments due to war or religious/cultural conflict. (f) before and (g) after destruction of Temple of Bel at Palmyra in 2015 (Syria).

(a)

(b)

Figure 6.73 Damage in the walls of the church of Santa Maria del Mar (Barcelona) caused by shell fragments and shrapnel during the War of the Spanish Succession, (a) with and (b) without growing vegetation.

monuments. This is the case of the Chapel of Scrovegni in Padua, Italy, frescoed by Giotto in 1303–1305 (Figure 6.75a). To limit the risk related to microclimate alterations (e.g. evaporation–condensation cycles, gaseous pollutants and daily variations of temperature), in 2000, the chapel was provided with an 'Equipped Technological Chamber', where the visitors wait for 15 min to allow their body humidity to be lowered and smog dusts to be filtered (Figure 6.75b). Inside the chapel, temperature and humidity are constantly monitored, and the visits are restricted to groups of maximum 25 people at a time. Examples of mass tourism are further given in Figure 6.76.

(b)

Figure 6.74 Modern living and urbanization. (a) Excavation for the metro line in front of the Arch of Constantine and Colosseum, Rome, during 1940–1950s and (b) urban skyscraper context for a masonry church, Christ Church Cathedral, Montreal.

(a)   (b)

Figure 6.75 Scrovegni Chapel in Padua. (a) Equipped Technological Chamber and (b) frescoes by Giotto between 1303 and 1305.

(a)

(b)

Figure 6.76 Examples of mass tourism. (a) Saint Peter's Basilica in Rome and (b) Great Wall of China.

Conclusions

'What does conservation mean?' This was the question that opened the preface of the present book. The first two chapters dealt with this topic, explaining its evolution, vicissitudes, successes and mistakes, up to the modern understanding. Centuries of debates have generated very different theories (e.g. from stylistic to critical restoration), all of which have contributed, to lesser or larger extent, to the modern understanding.

Experts accept today that every monument constitutes a 'genuine' problem and that no general or universal conservation and restoration methods or criteria exist. Therefore, on a case-by-case basis, the different stakeholders are called upon to propose interventions supported by a critical study of information sources, such as archaeology, history and science, but also considering aesthetics and the possibility of modern (but respectful) uses. In this wide and complex scenario, a remarkable role for the scientific basis of conservation and restoration is currently played by international organizations (namely ICOMOS – the International Council on Monuments and Sites) and documents (e.g. Venice Charter and ISCARSAH recommendations). ISCARSAH is the International Scientific Committee on the Analysis and Restoration of Structures of Architectural Heritage, founded by ICOMOS in 1996.

Only in recent times, the structure has been identified as a distinct part of the monument, a cultural wealth in itself, with precious documentary value. As a fundamental part in the 10,000-year-old history of construction, which coincides with the history of civilization itself, structure needs to be preserved as a valuable, but still alive, 'document' and testimony of our ancestors' structural insight and construction skills.

With this purpose, the book illustrated the modern/scientific approach to structural issues as recommended by ISCARSAH. By means of principles and guidelines, the study of heritage structures is based on a combination of different sources of qualitative and quantitative information. Using an analogy with the medical approach, the analyst should gather significant data (anamnesis), identify the causes of damage and decay (diagnosis), propose remedial measures (therapy) and monitor the efficiency of the interventions (control). All steps of the process are based on historical investigations, inspections, monitoring and structural analysis. However, and due to the important uncertainties and data limitations normally encountered in the study of ancient constructions, subjectivity is still possible. Due to it, experts' personal judgement is of critical importance.

With particular regard to heritage structures, Chapters 3 and 4 provided a comprehensive overview of the construction materials and techniques usually encountered in conservation projects. This knowledge is considered of fundamental importance in the anamnesis phase, as it can convey precious information about the essence of the construction and the possible causes of damage: a detailed (and non-conventional) structural assessment cannot disregard these aspects. Accordingly, due to their countless shapes and their fascinating role within the built cultural heritage, masonry vaults deserved a deeper discussion. Moreover, as getting part of the modern heritage, structural systems adopted in the recent past are also illustrated.

History of construction revealed the uninterrupted progress of ancient builders in achieving a high level of complexity and perfection. An analogously continuous effort was made in studying and explaining the statics of such complex elements. In ancient times, the design process followed what would be presently defined as 'a rudimentary scientific approach', that is, trial and error. Considering each building as a scaled specimen of a new one to be built, the ancient builders achieved a proper competence made up by simple geometrical criteria, gathered under the designation of *rules of thumb*. This almost forgotten knowledge, validated by the very existence of those buildings today, represents a historic and documentary value that must be preserved. Chapter 5 is devoted to introducing this topic.

On the other hand, besides a few isolated cases (among which the genius of Leonardo da Vinci), it is only at the end of the 17th century that the scientific community undertook a methodological study of the arch stability. Hooke, Bernoulli, Couplet, Coulomb and Mascheroni are just a few of the scholars who paved the way to the modern understanding of limit analysis. Given its importance for the intuitive comprehension and quick assessment of historic structure stability, the limit theory is briefly reviewed in the second part of Chapter 5.

Finally, Chapter 6 illustrated the main damages and collapse mechanisms in masonry buildings. Far from being exhaustive, several examples are proposed and the principal causes discussed. Understanding damage is a fundament part in forensics engineering. Without proper diagnostics, the safety assessment and the definition of remedial measures are, necessarily, incorrect. As in medicine, there is a need to address the causes and not the symptoms.

Finally, it is stressed that the topics covered in the present book are complex and broad, requiring time for maturity and, in many cases, further reading, reflection and experience. The authors hope that the reader can feel motivated and attracted by the growing field of conservation of heritage structures.

Bibliography

Abraham, P. (1934) *Viollet-le-Duc et le rationalisme médiéval*. Paris: Vicent, Fréal et Cie.

Addis, W. (2007) *Building: 3000 Years of Design Engineering and Construction*. London: Phaidon.

Alberti, L. B. (1485) *De re aedificatoria*, Edited by 1784 (Translated in Italian by Cosimo Bartoli, I Dieci Libri Di Architettura. Roma: Giovanni Zempel, Nicolai Laurentii Alamani.

Allen, E. and Iano, J. (2013) *Fundamentals of Building Construction: Materials and Methods*, 6th edn. Hoboken: John Wiley & Sons.

Andreu, A., Gil, L. and Roca, P. (2007) Computational analysis of masonry structures with a funicular model, Journal of Engineering Mechanics. American Society of Civil Engineers, 133(4), pp. 473–480. doi: 10.1061/(ASCE)0733-9399(2007)133:4(473).

Becchi, A. and Foce, F. (2002) *Degli archi e delle volte. Arte del costruire tra meccanica e stereotomia*. Venezia: Marsilio.

Benvenuto, E. (1991) *An Introduction to the History of Structural Mechanics - Part II: Vaulted Structures and Elastic*. New York: Springer-Verlag Inc.

Binda, L. (ed.) (2008) *Learning from Failure: Long-Term Behaviour of Heavy Masonry Structures*. Southampton: WIT Press.

Block, P. (2009) Thrust network analysis: Exploring three-dimensional equilibrium. PhD dissertation, Massachusetts Institute of Technology. Available at: http://hdl.handle.net/1721.1/49539.

Block, P., DeJong, M. and Ochsendorf, J. (2006) As hangs the flexible line: Equilibrium of masonry arches, *Nexus Network Journal*, 8(2), pp. 13–24. doi: 10.1007/s00004-006-0015-9.

Blondel, F. (1675) *Cours d'architecture enseigné dans l'Academie royale d'architecture*. Paris: Lambert Roulland.

Bussell, M. N. (1997) *Appraisal of Existing Iron and Steel Structures*. Ascot: Steel Construction Institute.

Carbonara, G. (1996) *Trattato di restauro architettonico*. Turin: UTET.

Ching, F. D. K. (1995) *A Visual Dictionary of Architecture*. New York: John Wiley & Sons.

Ciocci, M. P., Sharma, S. and Lourenço, P. B. (2018) Engineering simulations of a super-complex cultural heritage building: Ica Cathedral in Peru, *Meccanica*, 53(7), pp. 1931–1958. doi: 10.1007/s11012-017-0720-3.

Coenen, U. (1990) *Die spätgotischen Werkmeisterbücher in Deutschland. Untersuchung und Edition der Lehrschriften für Entwurf und Ausführung von Sakralbauten*. München: Scaneg.

Como, M. (2013) *Statics of Historic Masonry Constructions*. Berlin, Heidelberg: Springer (Springer Series in Solid and Structural Mechanics). ISBN: 978-3-642-30132-2.

Couplet, P. (1729) De la poussée des voûtes. In: *Mémoires de l'Académie Royale des Sciences de Paris*. pp. 79–117.

Couplet, P. (1730) Seconde partie de l'examen de la poussée des voûtes. In: *Mémoires de l'Académie Royale des Sciences Paris*, pp. 117.

Croci, G. (2008) *The Conservation and Structural Restoration of Architectural Heritage*. Southampton: WIT Press.

D'Ayala, D. and Casapulla, C. (2001) Limit state analysis of hemispherical domes with finite friction. In: Lourenço, P.B. and Roca, P. (eds) Structural Analysis of Historical Constructions. Guimarães, pp. 617–626.

Davey, N. (1961) *A History of Building Materials*. London: Phoenix House.

de Honnecourt, V. (1230) Sketchbook, https://gallica.bnf.fr/ark:/12148/btv1b10509412z/f36.image.r=villard%20de%20honnecourt.

Derand, F. (1643) *L'architecture des voûtes, ou l'art des traits et coupes des voûtes*. Paris: André Cailleau.

Di Pasquale, S. (1996) *L'arte del costruire. Tra conoscenza e scienza*. Venezia: Marsilio.

Esponda, M. (2003) Evolución de los criterios de intervención con hormigón armado en la restauración de edificios históricos en México y España. PhD dissertation, Universitat Politècnica de Catalunya, Barcelona.

Feilden, B. (2003) *Conservation of Historic Buildings*, 3rd edn. Oxford: Architectural Press, Elsevier.

Fernández Alba, A. et al. (1997) *Teoría e historia de la restauración*. Madrid: Munilla-Lería.

Frézier, A. F. (1737) *La théorie et la pratique de la coupe de pierres et des bois pour la construction des voûtes et autres parties des bâtiments civils et militaires, ou traité de stéréotomie à l'usage de l'architecture (1737–1739)*. Strasbourg/Paris: Charles-Antoine Jombert.

Gaetani, A. et al. (2016) Design and analysis of cross vaults along history, *International Journal of Architectural Heritage*, 10(7), pp. 841–856. doi: 10.1080/15583058.2015.1132020.

Giuffrè, A. (1993) *Sicurezza e conservazione dei centri storici. Il caso Ortigia*. Bari: Laterza.

Guerra Pestonit, R. A. (2012) Bóvedas y contrarresto del Colegio de Nuestra Señora de la Antigua de Monforte de Lemos : geometría, construcción y mecánica. PhD dissertation, Universidad Politécnica de Madrid (UPM).

Heyman, J. (1966) The stone skeleton, *International Journal of Solids and Structures*, 2(2), pp. 249–279. doi: 10.1016/0020-7683(66)90018-7.

Heyman, J. (1972a) *Coulomb's Memoir on Statics: An Essay in the History of Civil Engineering*. Cambridge: Cambridge University Press.

Heyman, J. (1972b) "Gothic" construction in Ancient Greece, *Journal of the Society of Architectural Historians*, 31(1), pp. 3–9. doi: 10.2307/988722.

Heyman, J. (1982) *The Masonry Arch*. Chichester: Ellis Horwood Ltd.

Heyman, J. (1983) Chronic defects in masonry vaults: Sabouret's cracks, *Monumentum*, 26(2), pp. 131–141.

Heyman, J. (1995) *The Stone Skeleton: Structural Engineering of Masonry Architecture*. Cambridge, UK: Cambridge University Press.

Huerta, S. (2004) *Arcos, bóvedas y cúpulas: geometría y equilibrio en el cálculo tradicional de estructuras de fábrica*. Madrid: Instituto Juan de Herrera.

Huerta, S. (2009) The debate about the structural behaviour of gothic vaults: From Viollet-le-Duc to Heyman. In: Kurrer, K.-E., Lorenz, W., and Wetz, V. (eds.) In *Proceedings of the Third International Congress on* Construction History, Cottbus, pp. 837–844. ISBN: 978-3-936033-31-1, www.researchgate.net/publication/41764961_The_Debate_about_the_Structural_Behaviour_of_Gothic_Vaults_From_Viollet-le-Duc_to_Heyman.

Huerta, S. (2010) Thomas Young's theory of the arch: Thermal effects. In: Sinopoli, A. (ed.) *Mechanics and Architecture: Between Epistéme and Téchne*. Roma: Edizioni di Storia e Letteratura, pp. 155–178.

ICOMOS (1964) The Venice Charter: International charter for the conservation and restoration of monuments and sites. *In Second International Congress of Architects and Technicians of Historic Buildings*, Venice.

ICOMOS (1994) Nara document on authenticity. *In Conference on Authenticity in Relation to the World Heritage Convention*, Nara, Japan.

ICOMOS (2008) Charter on the interpretation and presentation of cultural heritage sites. *Ratified by the 16th General Assembly of ICOMOS*, Quebec, Canada, 4th October 2008.

ICOMOS (2014) Approaches for the conservation of twentieth-century architectural heritage. Madrid document.

ICOMOS (2017) Guidance on post trauma recovery and reconstruction for world heritage cultural properties.

ICOMOS/ISCARSAH (2003) Recommendations for the analysis, conservation and structural restoration of architectural heritage.

ISO 13822:2010 Bases for design of structures - Assessment of existing structures. Annex I on Heritage Structures. (TC98, SC2, WG6), www.iso.org/standard/46556.html.

Jokilehto, J. (2018) *A History of Architectural Conservation*, 2nd edn. Abingdon: Routledge.

Kurrer, K.-E. (2008) *The History of the Theory of Structures: From Arch Analysis to Computational Mechanics*. Berlin: Ernst & Sohn.

Lourenço, P. B. (2001) Analysis of historical constructions: From thrust-lines to advanced simulations. In: Lourenço, P. B. and Roca, P. (eds) *Structural Analysis of Historical Constructions*. Guimarães: Universidade do Minho, pp. 91–116.

Lourenço, P. B. (2002) Computations on historic masonry structures, *Progress in Structural Engineering and Materials*, 4(3), pp. 301–319. doi: 10.1002/pse.120.

Lourenço, P. B. et al. (2007) Failure analysis of Monastery of Jerónimos, Lisbon: How to learn from sophisticated numerical models, *Engineering Failure Analysis*, 14(2), pp. 280–300. doi: 10.1016/j.engfailanal.2006.02.002.

Mainstone, R. (1998) *Developments in Structural Form*, 2nd edn. Oxford: Architectural Press.

Mark, R. (1982) *Experiments in Gothic Structure*. Cambridge: MIT Press.

Mark, R. (1990) *Light, Wind, and Structure: The Mystery of the Master Builders*, 1st edn. Cambridge (MA): MIT Press.

Mark, R. (1993) *Architectural Technology up to the Scientific Revolution: The Art and Structure of Large-Scale Buildings*, 1st edn. Cambridge (MA): MIT Press.

Mark, R., Abel, J. F. and O'Neill, K. (1973) Photoelastic and finite-element analysis of a quadripartite vault, *Experimental Mechanics*, 13(8), pp. 322–329. doi: 10.1007/BF02322391.

Maynou, J. (2001) Estudi estructural del pòrtic tipus de la Catedral de Mallorca mitjançan l'estàtica gràfica. Master thesis, Universitat Politècnica de Catalunya, Barcelona.

MIT - Philippe Block (2005) InteractiveTHRUST. Available at: http://web.mit.edu/masonry/interactiveThrust/index.html (Accessed: 20 December 2018).

Müller, W. (1990) *Grundlagen gotischer Bautechnik*. Munchen: D. Kunstverlag.

National Civil Protection Service (2013) In: Papa, S. and Pasquale, G. D. (eds) *Manuale per la compilazione della scheda per il rilievo del danno ai beni culturali, chiese - MODELLO A-DC*, www.protezionecivile.gov. it/en/transparent-administration/legal-measures/detail/-/asset_publisher/default/content/dpcm-del-13-marzo-2013-approvazione-del-manuale-per-compilare-la-scheda-di-rilievo-del-danno-ai-beni-culturali.

National Geographic (2014) How an amateur built the world's biggest dome. Available at: https://youtube.com/watch?v=_IOPlGPQPuM (Accessed: 19 December 2018).

Nova PBS (2015) Great cathedral mystery PBS. Available at: https://youtube.com/watch?v=_kCZeN6d1cA (Accessed: 19 December 2018).

O'Dwyer, D. (1999) Funicular analysis of masonry vaults, *Computers and Structures*, 73(1–5), pp. 187–197. doi: 10.1016/S0045-7949(98)00279-X.

O'Hearne, N., Mendes, N. and Lourenço, P. B. (2018) Seismic analysis of the San Sebastian Basilica (Philippines). *In Proceedings of 40th IABSE Symposium*. Nantes, France.

Ottoni, F. and Blasi, C. (2015) Results of a 60-year monitoring system for Santa Maria del Fiore Dome in Florence, *International Journal of Architectural Heritage*, 9(1), pp. 7–24. doi: 10.1080/15583058.2013.815291.

Pelà, L. et al. (2016) Analysis of the effect of provisional ties on the construction and current deformation of Mallorca Cathedral, *International Journal of Architectural Heritage*, 10(4), pp. 418–437. doi: 10.1080/15583058.2014.996920.

Poleni, G. (1748) *Memorie istoriche della gran cupola del tempio Vaticano e de' danni di essa, e de' ristoramenti loro*. Padua: Stamperia del Seminario.

Regione Marche, CNR and University of L'Aquila (2007) Repertorio dei meccanismi di danno, delle tecniche di intervento e dei relativi costi negli edifici in muratura. Decreto del Commissario Delegato per gli interventi di protezione civile n. 28 del 10 aprile 2002.

Regione Toscana (2003) *Istruzioni tecniche per l'interpretazione ed il rilievo per macroelementi del danno e della vulnerabilità sismica delle chiese*. Venzone: ARX s.c.r.l.

Roca, P. (2001) Studies on the structure of Gothic Cathedrals. In: Lourenço, P. B. and Roca, P. (eds) *Structural Analysis of Historical Constructions*. Guimarães: Universidade do Minho, pp. 71–90.

Roca, P. et al. (2008) Monitoring of long-term damage in long-span masonry constructions. In: Binda, L. (ed.) *Learning from Failure*. Southampton: WIT Press.

Roca, P. et al. (2010) Structural analysis of masonry historical constructions. Classical and advanced approaches, *Archives of Computational Methods in Engineering*, 17(3), pp. 299–325. doi: 10.1007/s11831-010-9046-1.

Roca, P. et al. (2013) Continuum FE models for the analysis of Mallorca Cathedral, *Engineering Structures*, 46, pp. 653–670. doi: 10.1016/j.engstruct.2012.08.005.

Rubió y Bellver, J. (1912) Conferencia acerca de los conceptos orgánicos, mecánicos y constructivos de la Catedral de Mallorca'. In *Anuario de la Asociación de Arquitectos de Cataluña*, Barcelona, pp. 87–140.

Sabouret, V. (1928) Les voûtes d'arêtes nervurées. Rôle simplement décoratif des nervures, *Le Génie Civil*, 92, pp. 205–209.

Schön, J. (2015) *Physical Properties of Rocks: Fundamentals and Principles of Petrophysics*. Amsterdam: Elsevier.

Serlio, S. (1619) *Tutte l'opere d'architettura et prospetiva*, 2nd edn. Venice: Appresso Giacomo de' Franceschi.

Siegesmund, S. and Snethlage, R. (eds) (2014) *Stone in Architecture: Properties, Durability*, 5th edn. Berlin, Heidelberg: Springer. ISBN: 978-3-642-45155-3.

Silva, L. C. et al. (2018) Seismic structural assessment of the Christchurch Catholic Basilica, New Zealand, *Structures*, 15, pp. 115–130. doi: 10.1016/j.istruc.2018.06.004.

Straub, H. (1992) *Die Geschichte der Bauingenieurkunst*, 4th edn. Basel: Birkhäuser.

Tampone, G. (1996) *Il restauro delle strutture di legno*. Milano: Hoepli.

Tosca, T. V. (1707) *Compendio mathemático en que se contienen todas las materias más principales de las ciencias que tratan de la cantidad (1707–15)*. Valencia: Antonio Bordazar. (2nd ed. Madrid, 1721–1727. Fifth volume reprinted in Arquitectura civil, montea y cantería: Valencia: Librería París-Valencia, 1992; Universidad Politécnica, 2000).

UNESCO (1972) Convention concerning the protection of the world cultural and natural heritage. Paris.

Ungewitter, G. G. and Mohrmann, K. (1890) *Lehrbuch der gotischen Konstruktionen*, 3rd edn. Leipzig: Weigel.

UNWTO (2017) *UNWTO Tourism Highlights: 2017 Edition*. World Tourism Organization (UNWTO). ISBN: 9789284419029, www.e-unwto.org/doi/pdf/10.18111/9789284419029.

Verstrynge, E. et al. (2011) Modelling and analysis of time-dependent behaviour of historical masonry under high stress levels, *Engineering Structures*, 33(1), pp. 210–217. doi: 10.1016/j.engstruct.2010.10.010.

Wolfe, W. S. (1921) *Graphical Analysis: A Handbook on Graphic Statics*. New York: McGraw-Hill Book Company.

Permissions

CHAPTER 1

Image	License	Right holder
Image 1.2b		Luis Ramos
Image 1.4b1		Guerra-Pestonit, Rosa Ana, PhD Thesis "Bóvedas y contrarresto del Colegio de Nuestra Señora de la Antigua de Monforte de Lemos: geometría, construcción y mecánica". Year of publication: 2012. Universidad Politécnica de Madrid
Image 1.4b2		Guerra-Pestonit, Rosa Ana, PhD Thesis "Bóvedas y contrarresto del Colegio de Nuestra Señora de la Antigua de Monforte de Lemos: geometría, construcción y mecánica". Year of publication: 2012. Universidad Politécnica de Madrid
Table 1.2.1		Claudio Falasca
Table 1.2.2	GNU v. 1.2	Hans Schneider
Table 1.2.3b		Roland Scheidemann/picture-alliance/dpa/IPA
Table 1.2.3c	Public domain	Ingersoll
Table 1.2.4	CC BY-SA 3.0	PetrusSilesius
Table 1.2.5b	CC BY-SA 3.0	gilibean
Table 1.3.9		Helder Sousa
Table 1.3.10		Luis Ramos
Table 1.6.2.4		Pilar Giráldez and Màrius Vendrell
Table 1.6.2.7		DGPC/SIPA FOTO. 00051681; Sé do Porto; s.a.; 1951
Table 1.7.3.16	CC BY-SA 2.0	willow wilcox fox
Image 1.6b	CC BY 2.0	Feliciano Guimarães
Image 1.7a		Document ACM LFA-1724, F.53R, courtesy of Archive of the Chapter of Mallorca Cathedral and the source

(Continued)

Image	License	Right holder
Image 1.7b		Courtesy of J. Domenge and Institut d'Estudis Baleàrics: "J. Domenge, L'obra de la seu. El procés de construcció de la catedral de Mallorca en el tres-cents, Palma: Institut d'Estudis Baleàrics, 1997"
Image 1.9a1		Giorgos Karanikoloudis
Image 1.9b1		Maria Pia Ciocci
Image 1.9b2		Maria Pia Ciocci
Image 1.10a	CC BY 3.0	Sailko
Image 1.10b	CC BY 3.0	Sailko
Image 1.11a	Public domain	AA.VV., *Gaudí 2002. Miscel·lània*, Ed. Planeta, Barcelona (2002), ISBN 84-9708-093-9
Image 1.11b	CC BY-ND 2.0	Eva
Image 1.12a		Courtesy of Robert Mark: Mark, R., Abel, J.F. and O'Neill, K., 1973. Photoelastic and finite-element analysis of a quadripartite vault. Experimental Mechanics, 13(8): 322–329
Image 1.12b		Courtesy of Robert Mark: Mark R., 1982. *Experiments in gothic structure*, Cambridge: MIT Press.
Image 1.13a		Alessandra Marotta
Image 1.13b		Luis Silva
Image 1.13d		Nuno Mendes
Image 1.13e		Giorgos Karanikoloudis
Image 1.13f		Maria Pia Ciocci
Table 1.8.1		Courtesy of Jan Kozak: Image-KZ103, *Kozak Private Collection, Czech Republic*
Table 1.8.4		Miquel Llorens

CHAPTER 2

Image	License	Right holder
Image 2.1a	CC BY-SA 3.0	Tamer Abdo
Image 2.1b	CC BY-SA 2.0	Dennis Jarvis
Image 2.3a		Rabia Sangun
Image 2.4a	CC BY-SA 3.0	Hans Peter Schaefer, http://www.reserv-a-rt.de
Image 2.4b	CC BY-SA 4.0	Andrea Frascari
Image 2.5a	CC0 1.0	Gisela huerta
Image 2.5b	CC BY-SA 3.0	Jebulon
Image 2.6b	CC BY-SA 1.0	J. Miers, Jtesla16 at wts wikivoyage
Image 2.7a		Bologna, Fototeca Zeri, Photo inventory nr. 122175_g
Image 2.7b	CC BY 3.0	Mikhail Malykh
Image 2.8a		Credit: "Bibliothèque nationale de France" or "BnF"

(Continued)

Image	License	Right holder
Image 2.9a	CC0 1.0	Gilman Collection, Purchase, Harriette and Noel Levine Gift, 2005
Image 2.9b	CC BY-SA 3.0	Philipp Hertzog
Image 2.10b	CC BY-SA 4.0	Alexicographie
Image 2.11a		Collection Matteo Cerizza
Image 2.11b		Collection Giorgio Stagni
Image 2.13b	CC BY-SA 4.0	Notafly2
Image 2.14a	CC0 1.0	AlfvanBeem
Image 2.14b	CC BY-SA 2.0	Carole Raddato
Image 2.15b		The 70th Infantry Division Association (http://www.trailblazersww2.org)
Image 2.15c	CC BY 3.0	Schlaier
Image 2.16b	CC BY 2.0	Shadowgate
Image 2.17a	Public domain	War Office official photographer, Taylor (Mr)
Image 2.17b	CC BY-SA 3.0	Andrew Walker (walker44)
Image 2.18a	Public domain	Ukjent
Image 2.18b	CC BY-SA 2.0	Dennis Jarvis

CHAPTER 3

Image	License	Right holder
Table 3.1d	CC BY-SA 3.0	Notafly
Table 3.1f	CC BY-SA 3.0	Wknight94
Table 3.1l	CC BY-SA 3.0	Camelia.boban
Table 3.1n	CC BY-SA 3.0	Camelia.boban
Image 3.6a	CC BY-SA 2.0	Dennis Jarvis
Image 3.6b	CC BY-SA 2.0	Dennis Jarvis
Image 3.8b	CC BY-SA 2.0	Scott Dexter
Image 3.9b		Pilar Giráldez and Màrius Vendrell
Image 3.15a	CC BY-SA 4.0	Wolfgang Sauber
Image 3.15b	CC BY-SA 4.0	Einsamer Schütze
Image 3.17a	CC BY-SA 3.0	663highland
Image 3.17b	CC BY-SA 3.0	663highland
Image 3.19	CC BY-SA 4.0	NikosFF
Image 3.21b		Courtesy of Lynn T. Courtenay (cover page of the Journal of the Timber Framers Guild, Nr. 72 / June 2004)
Image 3.26a	Public domain	J.M. Petzinger
Image 3.32b	CC BY-SA 2.0	Rapsak

(Continued)

Image	License	Right holder
Image 3.33a	CC BY 2.0	Patrick Denker from Athens, GA This file is licensed under the Creative Commons Attribution 2.0 Generic license.
Image 3.34	No known copyright restrictions	The British Library @ Flickr Commons
Image 3.35a	Public domain	Stanley P. Mixon
Image 3.37b	CC BY 2.0	Jean-Pol GRANDMONT
Image 3.38b	CC BY-SA 2.0	Richard Croft
Image 3.39a	CC BY-SA 4.0	Peter K Burian
Image 3.40a	CC BY-SA 4.0	Rijksdienst voor het Cultureel Erfgoed
Image 3.40b	CC BY 2.0	Paul Mannix
Image 3.41a	CC BY 3.0	Photo by and (c) 2007 David Chen
Image 3.41b	CC BY-SA 3.0	Zeus1234
Image 3.43b	CC BY-SA 3.0	Rick Fink Cliveden of the National Trust
Image 3.43c	CC BY-SA 4.0	Ing.Mgr.Jozef Kotulič
Image 3.43d		Photo Jiri Blaha, 2007
Image 3.44a	Public domain	Bequest of Richard B. Seager, 1926
Image 3.45a		British Museum EA6046
Image 3.45b	CC0 1.0	Rogers Fund, 1931
Image 3.46a	CC BY-SA 3.0	663highland
Image 3.46b	CC0 1.0	Rogers Fund, 1907
Image 3.48a	Public domain	n·e·r·g·a·l
Image 3.51d1	CC BY-SA 4.0	G41rn8
Image 3.51d2	CC BY-SA 4.0	G41rn8
Image 3.54b	CC BY-SA 2.0	Dennis Jarvis
Image 3.55a	CC BY-SA 4.0	Photo of ASI monument number N-OR-63 by Mano49j
Image 3.55b	CC BY-SA 4.0	Pratishkhedekar
Image 3.56a	CC BY-SA 2.0	Eurico Zimbres FGEL/UERJ
Image 3.56b	CC0 1.0	Borvan53
Image 3.59a	CC BY-SA 3.0	EmDee
Image 3.59b	Public domain	Pearson Scott Foresman
Image 3.60b	CC-BY-SA	© Max Planck Institute for the History of Science, Library
Image 3.61b	Public domain	Library of Congress Prints and Photographs Division Washington, D.C., HAER ILL,81-ROCIL,3-68-6
Image 3.62a		© Crown copyright: Royal Commission on the Ancient and Historical Monuments of Wales © Hawlfraint y Goron: Comisiwn Brenhinol Henebion Cymru

(Continued)

Image	License	Right holder
Image 3.62b		© Crown copyright: Royal Commission on the Ancient and Historical Monuments of Wales © Hawlfraint y Goron: Comisiwn Brenhinol Henebion Cymru
Image 3.65b		Maria Isabel Brito Valente
Image 3.65c		Maria Isabel Brito Valente
Image 3.67a	CC BY 3.0	stavros1
Image 3.67b	Public domain	Photo by Eusebius (Guillaume Piolle)
Image 3.68a	CC BY 2.0	Jules & Jenny
Image 3.68b	CC BY-SA 4.0	Shitha Valsan
Image 3.70a		Maria Giovanna Masciotta
Image 3.70b		Maria Giovanna Masciotta
Image 3.70c	CC BY-SA 3.0	Karl Gruber
Image 3.71a	CC BY 3.0	JCNazza
Image 3.71b	CC BY-SA 2.0	Akke
Image 3.71c	CC BY 3.0	Ralf Houven
Image 3.72a	CC BY 2.0	Marie-Lan Nguyen
Image 3.72b	No known copyright restrictions	Internet Archive Book Images
Image 3.73a	CC BY 2.0	bvi4092
Image 3.73b	CC BY-SA 4.0	Stephjeb
Image 3.73c	CC BY 4.0	Wellcome Images, ICV No 24789, Photo number: V0024348
Image 3.73d	Public domain	User:Velela
Image 3.73e	CC BY-SA 2.0	Gimli_36
Image 3.73f	CC BY-SA 3.0	MaryDo
Image 3.74b	CC BY-SA 2.0	Dennis Jarvis
Image 3.75a	CC BY 2.0	Jeff Kopp
Image 3.75c	CC BY-SA 4.0	Monster4711
Image 3.76a	CC BY-SA 4.0	Paul Louis
Image 3.76b	CC0 1.0	Horta Museum
Image 3.77a	Public domain	Irving Underhill

CHAPTER 4

Image	License	Right holder
Image 4.7b	CC BY-SA 3.0	MOSSOT
Image 4.10b	CC BY-SA 3.0	Sailko
Image 4.10d	CC BY-SA 4.0	Shakko
Image 4.11c		Photo by Daniele Indrigo

(Continued)

Image	License	Right holder
Image 4.14a	CC BY-SA 3.0	Robert Freeman
Image 4.15a	CC BY-SA 2.0	Karen Green
Image 4.15b	Public domain	Charles Chipiez
Image 4.17a		Dorling Kindersley images (20213145)
Image 4.17b	Public domain	New York Public Library, Image ID 1542753
Image 4.18a1	CC BY-SA 4.0	AlejandroLinaresGarcia
Image 4.18b1	CC BY-SA 3.0	Laslovarga
Image 4.19a	CC BY-SA 4.0	Image by Luigi Tarascio - aeroshot.it
Image 4.19b	CC BY-SA 3.0	Miguel Hermoso Cuesta
Image 4.20a	CC BY-SA 3.0	Agnete
Image 4.21b	CC BY-SA 4.0	Xosema
Image 4.22a	CC-BY 2.5	Maros M r a z (Maros)
Image 4.22b1	Public domain	Penn State University Libraries @ Flickr
Image 4.22b2	Public domain	Georg Dehio/Gustav von Bezold
Image 4.23a	Public domain	Etienne Duperac
Image 4.23b	CC BY-SA 4.0	Dietmar Rabich
Image 4.25a	CC BY-SA 3.0	Bollweevil
Image 4.25b	CC BY-SA 3.0	Bollweevil
Image 4.26a	CC BY-SA 3.0	Adli Wahid
Image 4.26b		Rabia Sengun
Image 4.26c	CC BY-SA 3.0	Ogodej
Image 4.28	CC BY-SA 4.0	Rijksdienst voor het Cultureel Erfgoed
Image 4.30a	CC BY-SA 4.0	Enfo
Image 4.30b	CC BY 4.0	José Luis Filpo Cabana
Image 4.31a	CC BY-SA 3.0	Foto Fitti
Image 4.31b	CC BY-SA 3.0	Foto Fitti
Image 4.33a		© The British Library Board, B 1240 (390)
Image 4.33b		© The British Library Board, B 1240 (2046)
Image 4.34a	CC BY-SA 3.0	Myrabella
Image 4.34b		Gursoy Group Restoration
Image 4.35a	CC BY 3.0	Sinan Şahin
Image 4.35b	CC BY-SA 3.0	ROFI44WIK
Image 4.36a	CC BY-SA 4.0	Milafirenze
Image 4.36b	Public domain	Charles Herbert Moore
Image 4.37a		Museo Galileo, Firenze - Archivio Fotografico
Image 4.37b		Museo Galileo, Firenze - Archivio Fotografico
Image 4.37c	CC BY-SA 4.0	Peter K Burian
Image 4.38a	CC BY-SA 3.0	Jebulon
Image 4.38b	CC BY-SA 3.0	Ramon Espiña Fernand….
Image 4.40a		Ali Tavakoli Dinani

(Continued)

Image	License	Right holder
Image 4.40b		Ali Tavakoli Dinani
Image 4.40c		Ali Tavakoli Dinani
Image 4.41a	CC BY-SA 3.0	Bernard Gagnon
Image 4.42b	Public domain	Thomas Ustick Walter
Image 4.43a	CC BY 3.0	Mister No
Image 4.43b	CC0 1.0	Marco Del Torchio
Image 4.44b	CC BY-SA 4.0	Txllxt TxllxT
Image 4.45a	CC BY-SA 3.0	Glabb
Image 4.45b	CC BY-SA 4.0	Giorgio L. Rutigliano
Image 4.46a	Public domain	Zoran Knez
Image 4.46b	Public domain	Joseolgon
Image 4.48	CC BY-SA 3.0	Rrm998
Image 4.49a	CC BY 2.0	Thelastminute (Duncan Rawlinson)
Image 4.49b	CC BY 2.0	juice.springsteen
Image 4.51a		City of Tacoma
Image 4.51b	CC BY-SA 2.0	SounderBruce
Image 4.53a		Société Freyssinet
Image 4.53b		Société Freyssinet
Image 4.55a	CC BY-SA 2.0	Guillaume Cattiaux
Image 4.55b	CC BY 2.0	Asp Explorer
Image 4.56a	CC BY-SA 3.0	Dge
Image 4.56b	CC BY-SA 2.0	Sergio Calleja (Life is a trip)
Image 4.56c	Public domain	GAED
Image 4.57a	CC BY-SA 4.0	ETH-Bibliothek Zürich, Bildarchiv / Fotograf: Comet Photo AG (Zürich) / Com_F68-13954
Image 4.57b	CC BY-SA 3.0	Хрюша
Image 4.60a	CC BY-SA 4.0	Outisnn
Image 4.61a		Courtesy Pier Luigi Nervi Project Association, Brussels
Image 4.61b		Courtesy Pier Luigi Nervi Project Association, Brussels
Image 4.62a		Courtesy Pier Luigi Nervi Project Association, Brussels
Image 4.62b		Courtesy Pier Luigi Nervi Project Association, Brussels
Image 4.63	CC BY 2.0	Alex Proimos
Image 4.64a	CC BY 3.0	JoJan
Image 4.64b	CC BY-SA 4.0	Groninger3
Image 4.65a	CC BY-SA 4.0	Camilla Vitoria Machado
Image 4.65b	CC BY-SA 3.0	jjandson

(Continued)

Image	License	Right holder
Image 4.66a	CC BY-SA 4.0	SymphonicPoet
Image 4.66b	CC BY 3.0	Bohao Zhao
Image 4.68a	CC BY-SA 3.0	Leah Rucker
Image 4.68b	CC BY 2.0	FRC® Team 836 The RoboBees
Image 4.69a		ANSA/AP Photo/Carl Perutz
Image 4.69b	CC BY 2.0	Rich Mitchell
Image 4.70a		Courtesy of TML
Image 4.70b		Courtesy of TML
Image 4.73a	CC BY-SA 4.0	Guerinf
Image 4.73b	CC BY-SA 3.0	Jorge Royan
Image 4.75a		© Rubner Holzbau
Image 4.75b		© Rubner Holzbau
Image 4.78b1 Image 4.78b2 Image 4.78b3		©VSL International
Image 4.81a		Julius Natterer
Image 4.81b		© Rubner Holzbau
Image 4.82a	CC BY-SA 3.0	Pere López Brosa
Image 4.82b	CC BY-SA 3.0	Katonams
Image 4.86		Climent Molins
Image 4.89a	CC BY-SA 3.0	Oderik
Image 4.89b	Public domain	Niplos
Image 4.90a	CC BY-ND 2.0	faungg's photos
Image 4.90b	Public domain	J Bar
Image 4.91a	No known copyright restrictions	Preus museum, Shot by Lewis Wickes Hine
Image 4.91b	Public domain	U.S. National Archives and Records Administration, Shot by Lewis Wickes Hine

CHAPTER 5

Image	License	Right holder
Image 5.5	CC BY-SA 4.0	Jordiferrer

CHAPTER 6

Image	License	Right holder
Image 6.2a		CR26/86/10 reproduced by kind permission of TRL
Image 6.4b		Daniel Oliveira

(Continued)

Image	License	Right holder
Image 6.9a		Giorgos Karanikoloudis
Image 6.16a		Michele Castobello
Image 6.16b		Michele Castobello
Image 6.16c		Michele Castobello
Image 6.17a		Verstrynge, E. and Van Gemert, D. (2018) 'Creep failure of two historical masonry towers: analysis from material to structure', Int. J. Masonry Research and Innovation, Vol. 3, No. 1, pp.50–71. (page 66, Fig. 14b) © Inderscience
Image 6.17b	CC BY-SA 4.0	Sally V
Image 6.20b		Pilar Giráldez and Màrius Vendrell
Image 6.23b		Pilar Giráldez and Màrius Vendrell
Image 6.40b1		Pilar Giráldez and Màrius Vendrell
Image 6.40b2		Pilar Giráldez and Màrius Vendrell
Image 6.42a	CC BY-SA 4.0	Enric
Image 6.42b		Ferran Vizoso
Image 6.45a	CC BY-SA 2.0	Richard Webb
Image 6.47a	CC BY 2.0	tronics
Image 6.47b	CC BY-SA 2.0	Eli Duke
Image 6.47c	CC BY-SA 3.0	KlausFoehl
Image 6.47d	CC BY 2.0	The Real Bear
Image 6.49	CC BY-SA 4.0	Mtaylor848
Image 6.50a	CC BY 3.0	Olaf Tausch
Image 6.50b	No known copyright restrictions	Cornell University Library @ Flickr Commons
Image 6.51b1		Luis Ramos
Image 6.51b2		Luis Ramos
Image 6.52c		DGPC/SIPA FOTO. 00051681; Sé do Porto; s.a.; 1951
Image 6.53a		Valeria Chiatti
Image 6.53b		Ottavio Abramo
Image 6.53c		Michele Ghisolfi
Image 6.54a	CC BY 2.0	Charlie Phillips
Image 6.54b	CC BY-SA	Diego Delso, delso.photo, License CC-BY-SA
Image 6.56a		Pilar Giráldez and Màrius Vendrell
Image 6.56b		Pilar Giráldez and Màrius Vendrell
Image 6.57a		Pilar Giráldez and Màrius Vendrell
Image 6.57b		Pilar Giráldez and Màrius Vendrell
Image 6.58a	CC BY-SA 3.0	Antonio Lara Muñoz

(Continued)

Image	License	Right holder
Image 6.58b		Pilar Giráldez and Màrius Vendrell
Image 6.60a		Pilar Giráldez and Màrius Vendrell
Image 6.61a		Maria Manuel Lobo Pinto Oliveira
Image 6.61b	CC BY-SA 4.0	Joseolgon
Image 6.63a	CC BY-SA 2.0	Dennis Jarvis
Image 6.63b	CC BY-SA 2.0	Photo © 2004 Jacob Rus
Image 6.64a		Pilar Giráldez and Màrius Vendrell
Image 6.64b		Pilar Giráldez and Màrius Vendrell
Image 6.65a	CC BY-SA 2.0	DAVID HOLT from London, England
Image 6.65b	CC BY-SA 3.0	Jaime Silva
Image 6.66b		*From the collection of the Local History & Genealogy Division, Rochester (NY) Public Library*
Image 6.67a		Bernard Bastien / SPIQ.ca
Image 6.67b	CC BY 3.0	Gilbert Bochenek
Image 6.68a	CC BY-SA 4.0	Felipe Milanez
Image 6.68b	CC BY-SA 4.0	Lu Brito
Image 6.69a	CC BY-SA 4.0	Marind
Image 6.69b	CC BY-SA 4.0	Zuffe & Louis H. G.
Image 6.72a	Public domain	UNESCO/A Lezine
Image 6.72b	CC BY 2.0	Carl Montgomery
Image 6.72d	CC BY 3.0	Donar Reiskoffer
Image 6.72e	CC BY-SA 3.0	Fer.filol
Image 6.72f	CC BY-SA 3.0	Jerzy Strzelecki
Image 6.72g	CC BY 4.0	Jawad Shaar, Tasnim News Agency
Image 6.73a		Pilar Giráldez and Màrius Vendrell
Image 6.73b		Pilar Giráldez and Màrius Vendrell
Image 6.74a	Public domain	Italia ENIT March 1941
Image 6.75a		Amira Al Habash
Image 6.75b	CC BY-SA 4.0	Derbrauni
Image 6.76a	CC BY-SA 3.0	Bjørn Christian Tørrissen
Image 6.76b	CC BY-SA 3.0	Asadal

Index

Index of Monuments

Index of Scholars